Our Fictional Minds

Moving beyond Consciousness, Self, and Other Illusions

David C. Fisher, Ph.D.

Prometheus Books

Essex, Connecticut

Prometheus Books

An imprint of The Globe Pequot Publishing Group, Inc.
64 South Main Street
Essex, CT 06426
www.globepequot.com

Distributed by NATIONAL BOOK NETWORK

British Library Cataloguing in Publication Information Available

Library of Congress Cataloging-in-Publication Data

Names: Fisher, David, 1958– author.
Title: Our fictional minds : moving beyond consciousness, self, and other illusions / David C. Fisher.
Description: Lanham, MD : Prometheus Books, [2024] | Includes bibliographical references. | Summary: "Our Fictional Minds examines and challenges our most common—and seemingly common-sense—ideas about human consciousness. Drawing on developments in neuroscience, psychology, and monitoring technology, psychologist David C. Fisher shows how and why our usual takes on the human mind both serve us and limit us"—Provided by publisher.
Identifiers: LCCN 2024012038 (print) | LCCN 2024012039 (ebook) | ISBN 9781493085330 (paperback) | ISBN 9781493085347 (epub)
Subjects: LCSH: Consciousness. | Brain.
Classification: LCC BF311 .F486 2024 (print) | LCC BF311 (ebook) | DDC 153—dc23/eng/20240603
LC record available at https://lccn.loc.gov/2024012038
LC ebook record available at https://lccn.loc.gov/2024012039

For my daughters Laura and Alissa:
I hope you continue to thrive by seeking your own answers
rather than by taking what I say too seriously.

CONTENTS

AUTHOR'S NOTE

Writing this book has been a fascinating walk, filled with personal struggles. Every chapter reflects a small part of my ever-changing self. Each one is a piece of the puzzle I've been putting together for decades.

When I began my career, I held many notions about human nature, often absorbed from family, friends, and culture. These ideas were foundational to my worldview. What I encountered later was often jarring. Letting go of some of my early thinking was difficult, but doing so introduced me to new adventures full of intrigue and excitement.

My patients have often been my best teachers. They have shown me that my learning was never complete. Seeing this helped keep my perspectives fresh, both at work and in my personal life.

This book is partly autobiographical, but it's not a memoir. Instead, I intend to use parts of my life to show the transformations possible when we continuously question and change our perspectives, guided by the wisdom of diverse philosophies and the people around us.

Part I

OUR ILLUSIONS—HOW THEY HELP AND HURT US

Things are not always what they seem; the first appearance deceives many.

—Phaedrus

CHAPTER ONE

ANNA AND HER THERAPIST

Anna, a strong and determined woman, loved to work in her garden. Through her many books, she enjoyed discovering new ideas. But she had a problem, one that had challenged her since childhood. She desperately wanted to lose weight. Her attempts to drop even a few pounds were full of setbacks. Diets and weight-loss programs gave her no lasting success. None of them fulfilled their promise of helping her feel better. She continued to wear the same size clothes and shied away from meeting new people. Although she was usually resilient, lasting change seemed elusive.*

But Anna was also adventurous, so one day she tried something new. She saw an experienced counselor, Antonio, who regularly helped patients with weight loss.

In his professional work, Antonio had seen his many patients' successes and failures. Fortunately, being adept at different weight-loss strategies, he offered several options for his patients. These skills, along with a wonderful ability to connect, made him an ideal match for Anna.

Antonio had recently tried hypnosis with his patients. Being naturally curious, he hoped it would help him uncover his patients' unconscious turmoil that was blocking their capacity for deep psychological change. He enjoyed that feeling of exploration and was fascinated by the stories his patients said they had uncovered about their pasts.

At first, Antonio wasn't sure how best to help Anna. He knew that everyone was unique and that what would work for her was not necessarily what helped others. And so he asked her how she had previously tried to lose weight. He also learned about her overall persistence by asking how she had overcome other obstacles in the past. Remembering that the novel prospect

---

* Of course, in this description, all information that would reveal Anna's identity has been changed or omitted.

of hypnosis had frightened some of his previous patients, Antonio was also trying to gauge her willingness to try new things.

In short, he was trying to put many pieces together and get a clear picture of Anna's personality and life experiences. He hoped these would help him devise a tailored approach that would enhance her chances for success.

As he listened to her, he soon came to believe that she would be an especially receptive hypnotic patient. His conclusion was strengthened after he saw her gaze follow his own when he glanced around the room. She also responded well to slight inflections in his voice. Seeing this, Antonio realized she was deeply engaged in a working relationship with him, even mirroring some of the most subtle changes in his behavior.[*]

Antonio gave Anna a thoughtful look. "I think hypnosis could help you reach your weight-loss goal," he said. "Do you mind if I give you a brief test? It will help me gauge your hypnotizability."

"Sure," Anna replied.

Antonio gave her several imaginative hypnotic prompts. First, he asked her to picture a fly landing on her cheek and to brush it off. Over the next few minutes, Antonio gave her more unusual suggestions. Then his suggestions grew even more challenging. He asked her to picture and hear a car inside his office. Next, he got her to confirm that she'd come directly to his office from her home, despite her bringing a bag of freshly bought groceries to the appointment. Antonio used a "hypnotizability scale" to tally her responses to some of his suggestions and then shared that she probably would be a good hypnotic subject. "Want to give hypnosis a try?" he asked.

"I'm in," Anna said.

Antonio believed that hypnosis would work better for her than traditional counseling, often called *talk therapy*. He thought it could reveal some of Anna's undiscovered thoughts, impulses, and motivations. Antonio reasoned that some destructive memories and beliefs were lurking in unexplored parts of her unconscious mind. By bringing these thoughts to light, he believed that Anna might clarify the reason for her overeating.

Feeling anxious, Anna asked if hypnosis would give Antonio control over her mind. He smiled. "Your mind is your brain.[†] It contains your thought processes, your emotions, wishes, perceptions, and reasoning. I can't

---

* While these behaviors suggested a course of action for Antonio, they should not be viewed as predictors of Anna's tendency to show hypnotic behaviors.

† As we will see later, not all follow Antonio's assessment of the nature of mind.

control any of those. Hypnosis won't cause you to lose control over anything. It won't hijack your brain. And it won't make you do anything that conflicts with your morals or cause you any other harm. But hopefully it will help us discover what makes you overeat."

Using the deliberate ceremony of an induction—including a lengthy suggestion in which Antonio told Anna that he would put her into a trance—he led her into a hypnotic state.* Antonio instructed her to relax, to feel her eyelids become heavy, and then to notice them shut, as if they had effortlessly closed on their own.

Antonio told Anna, "You'll soon see a long staircase going downstairs into another room. Please slowly and gently walk down it. Let your feet sink into the thick plush carpet and look at the pictures on the walls as you go." He was silent for a few seconds. "What do you see?" he asked.

Anna said, "I like the art on the walls. And the handrail is smooth and shiny, like it's been freshly varnished."

When she reached the bottom of the stairs, she said, "There's a room over there. It's beautiful!"

"Do you want to go inside?" Antonio asked.

"Yes. Yes, I do."

"Go ahead. And, once you're inside, I'm going to guide you in making the room pleasant and comfortable. You can turn it into any haven you'd like."

Anna saw a magnificent door which, at Antonio's urging, she opened. She slowly entered the room and, with Antonio's guidance, she crafted her own ideal place of refuge from the external chaos of her life. She mentally painted it with her favorite bright colors. She saw a beautiful light and even felt a pleasant cool breeze flowing over her arms. Soft music played in the background, and the scent of lilacs wafted by her. She settled on one of the thick cushions scattered about on the floor.

After giving her plenty of time to enjoy and relax, Antonio said, "All right, that's enough for now. You'll be able to revisit this room in our future sessions. Go ahead and walk back up the steps, and we'll finish for today."

During each of their next few meetings, Anna revisited this cherished room. Sometimes her friends were there, and she had enjoyable conversations

---

* An induction often has three parts. First, the subject is told that they will be given suggestions for changes in their thoughts or sensations. Second, they are helped to relax. Third, the hypnotist encourages the desired changes in the person's thoughts or sensations.

with them about food, theater, and her aspirations for her body. Most of the time, though, she chose to be alone in the safe and comfortable place she had created.

After her third visit to this room, Anna began to view hypnosis as a gift, one that helped her feel good. She was no longer afraid that hypnosis would make her lose control. To her, it now could lead to a healthier and more attractive version of herself. The hypnosis was, to her surprise, an unexpected ally.

During Anna's fifth visit to her private sanctuary, Antonio asked her in a relaxed, pleasant tone, "Anna, why do you eat too much?"

Anna paused for a few seconds, then said, "I'm not sure."

Antonio wondered if there was another part of Anna that knew why she overate. He said, "There must be someone else inside of you who knows the reason."*

For a moment, Anna said nothing. Then, in a youthful voice, there was a reply: "It's because of what happened to Anna when she was young."

Antonio said, "Thanks. That's helpful. With whom am I speaking?"

"Sarah."

"Hello, Sarah. I'm Antonio. Can you tell me more about Anna's problem?"

"Uh-huh. A long time ago, when Anna was just a grade schooler, Anna's classmates teased her because she was tall. She looked much different than her friends, and that led to her getting teased. They thought that she could be ridiculed and they wouldn't get into trouble. Before long, almost the whole school was talking. She only had a few friends who really stood by her.

When the ridicule was at its worst, Anna coped by finding comfort in food, and she gained weight. Every meal gave her a temporary escape from her peers' judgments. But Anna can't talk about her anger toward those brats. She buried it with silence. She was afraid of the other kids, even though she doesn't say so when she's awake. To this day, whenever she's upset, she eats."

---

* The therapeutic sessions described in this chapter utilized a technique called *ego state therapy* (Watkins, 1993). This therapeutic technique, in which a therapist encourages their client to create a seemingly separate—and fictional—personality, is now rarely used. More often, therapists today would help Anna through other means, such as by encouraging her to think of herself feeling healthy after changing her eating and exercise habits.

Antonio nodded and appreciated the honesty. "Sarah," he said, "Anna is grown up now—and she's not especially tall for her age. No one is going to tease her about her height anymore. Do you think you can get Anna to confront her fears? Do you think that will help her lose weight?"

Sarah nodded.

"Would you be willing to help her now?"

"Um . . . okay."

"Go ahead and share whatever you think will help her."

"All right . . . Anna, you don't need to be afraid of those kids. What they say can't hurt you unless you let it. You are a wonderful person and are much more than your weight. Hold your head up, remember all your wonderful qualities, and don't give anyone satisfaction by making you suffer. And when you feel better about yourself, tell people you're not afraid of anyone!"

The voice shifted back into Anna's normal tone—but it was strong. "Listen, all of you. You shouldn't have said that stuff to me. And you know what else? I'm done listening to what you say!"

"Good," said Antonio. "How do you feel now, Anna?"

After a pause, Anna said, "Relieved."

"Well done. Now, please thank Sarah for her help."

Anna did as Antonio suggested.

"Now, slowly, open your eyes and come out of the trance."

As Anna departed Antonio's office, she mentally relived the events of the past 50 minutes. She remembered her trance vividly. She was amazed that she had recalled the childhood trauma that she thought she had so long ago forgotten—or had hidden inside her mind. She was also astonished that, briefly, she had felt as if she were two different people at once.

Anna now felt as if Sarah had "merged" back into her. Anna acted like one person again instead of two. While there once seemed like there was a chasm between her two "personalities," now there was none. It was as if two separate streams unified into one river. Her healing felt monumental.

After that session, Anna began to shed weight. As the weeks passed, she lost over 30 pounds—and kept them off. She was convinced that the hypnosis had been successful.

But that is not the takeaway from this story.

After her success, Anna ran into her two best friends from grade school. She told them that, with a therapist's help, she had remembered the ridicule she had suffered. She thanked them for the support they had offered her during those difficult times.

Her friends were amazed, but not because of her weight loss. They were stunned after they agreed that they not only never gave her such support, but also that Anna *had never been ridiculed for having been tall.*

The supposed childhood cause of her overeating was made up. The people she thought had teased her never existed. Her friends' recollections also made it clear to her that her eating problem, which didn't start until she was a young adult, probably wasn't linked with childhood trauma.

Anna started to question her own stability. And so, she scheduled another appointment with Antonio. After hearing her concerns, he told her that nobody knows everything about the brain and that sometimes it surprised even him. But he said that her brain was incredibly powerful and that he was now learning that the brain was especially adept with its ability to make up different realities, even if they weren't true. He said that "other people also get things wrong, but they are still sane. Even some witnesses to crimes think they are telling the truth, but their memories are still wrong."

After considerable thought and personal struggle, Anna eventually recognized this for herself.

Nevertheless, she concluded that the therapy—and her own fabricated explanation for her overeating—had produced the results she desired.[*]

Accounting for Anna's Change

You might think that Antonio pressured Anna into knowingly lying about her experiences. But under hypnosis, Anna believed what she saw and felt. What she experienced was as real to her as her surroundings when her eyes were open.

But Antonio's guidance did, of course, steer her in another sense. His statement "There must be someone else inside of you who knows the reason" conveyed his expectation: manufacture another person who did have answers about why you overeat. He underscored this expectation with his reply "With whom am I speaking?" Clearly he wanted her to think of herself as if she were two people.

After this, Antonio had a realization. He now saw that he wasn't uncovering and then addressing recollections of trauma. He was not the

---

[*] Most therapists today consider this now-seldom-used therapy to be ill-advised with most patients.

archeologist, digging for long-forgotten memories, as he had thought. Basically, his supposed explorations into the depths of her unconscious mind were not what he had first considered.

Instead, he now saw that his statements encouraged Anna to *create* a reality, a fresh perspective on things, and one that he imagined at least had a chance of helping her to lose weight. He thought he was perhaps a catalyst for change, but not in the way he had first imagined.

This new clarity helped guide some of Antonio's work with other patients. He wisely became more reluctant to encourage other people to remember events that were not true. He now saw that *there was danger in that approach*. After all, if he could easily cause false memories of childhood teasing, some therapists could create untrue memories of other types of abuse. And so, like Anna, he gained insight into the immense power, and flexibility, of people's imaginations. In short, he grew as a professional.

No one knows exactly why Anna was able to lose weight. Perhaps the hypnotic trances helped something to shift inside her, even if the story she embraced about her childhood was untrue. Possibly, because she believed the story for a time, it gave her more confidence to focus on losing weight. Maybe she lost weight to please Antonio so he would feel that he was helping her. Once Antonio even wondered if "Sarah" had been an abstract representation of a deeper emotional issue, one that resolved with Anna's exploration. Or perhaps her heaviness was the result of an undiagnosed medical condition that coincidentally resolved itself at an opportune time.

Fortunately, she accepted the ambiguity surrounding the reasons for her weight loss. She was happy that she had success and had a new understanding about herself and the world.

Hypnosis had other unexpected benefits for Anna. After discussions with Antonio about her fluid perceptions of herself, she now understood a bigger and more profound issue: Her beliefs, correct or not, influenced her reality. With this revelation, she became a more curious, questioning individual. This characteristic helped her to be more open to novel ideas. She embraced the idea that perceptions of reality were not fixed but were fluid and often improved over time. Thus, she unlocked even more possibilities for her life.

And so, during hypnosis, the layers of Anna's mind unfolded. With this, she glimpsed her mind's intricacies, each having led to new insights about her struggles and how she viewed herself.

The Misleading Power of Manufactured Memories

Anna's meetings with her therapist were hardly unique. Under hypnosis, many people have thought they encountered younger versions of themselves, recalled having been abducted by aliens, or remembered a host of long-forgotten friends and other imaginary characters.

And hypnosis is not the only situation under which these experiences can occur. For example, another similar method used by a very small number of therapists is called *past-lives therapy* or *past-life regression*. In this therapy, hypnotized people are told to bring back "memories" of supposed past incarnations. The idea is that confronting long-ago conflicts from other lifetimes will help patients address their current problems. These patients sometimes also retreat to comfortable imaginary rooms, appear to converse with old friends, and seem to re-acquire personalities from long-forgotten lifetimes.

People following certain cultural or religious traditions, such as Buddhism, sometimes believe that they remember past lives.[1] Many believe that they have improved morally from one life to the next.

Thinking about past lives is interesting and alluring, isn't it? I've seen some people happily report that they once had impressive positions, such as being princes or physicians. But I haven't met more than one North American who thought they had been a janitor.

Although it may be tempting to think of ourselves with such grand pasts, the scientific skeptic within many of us might well doubt the idea of having them. So why do people continue to believe? I can't rule out that we don't have past lives. However, people from my own culture get some comfort in thinking that we have souls with a rich history. Maybe that helps us feel larger than life or, in a sense, even that we are immortal.

Past-life therapy often contains themes of continuity, redemption, and interconnectedness. Seeming to remember them could, in fact, sometimes be beneficial. We should therefore respect people who report them and remember that such memories cater to an almost universal human wish that we last longer than a single lifetime.

The revelation that false memories can readily appear—and in some cases seem to be helpful—can be unsettling. But all of us, at times, embrace thoughts even when we know—or at least suspect—that they aren't true.

Thinking that we are smart, witty, and good looking can help us be confident and assertive, even when we're none of those things. Believing that there is a spirit who supports us can comfort us and encourage us to persist. And finally, during war, promising soldiers that they are fighting for the divine might encourage them to believe that the fight is for a just purpose.

As this book emphasizes, many of our deeply held convictions about ourselves and the world do not match reality. Yet, as you will see, in many cases they nevertheless appear to positively shape our lives.

Mental Models: Complexity and Utility

Anna's initial view of herself—as one personality inside a body—was a *model*, a useful way to talk about her own actions and motivations.

Very simply, in this book a model can be understood as a way to talk about something by explaining it using other ideas we already know. For instance, we talk about rainbows in terms of the idea of color.

Models of reality are necessary to help us make decisions. Furthermore, they shape our perceptions and bridge gaps between what we already know and novel concepts. Also, in this text, you can think of words such as *concept*, *idea*, *theory*, and even *belief* as being (more or less) synonymous with *model*.

Mental models help us explain why we like certain people, study philosophy, become dentists, believe in an afterlife, or think that we freely control our own actions. For Anna, the model of being one person first helped her interpret her experiences. It also made it possible for her to communicate her thoughts and needs to others in a way to which others could relate.

As we saw, Anna's model of only one person being inside her head shifted after Antonio's suggestions. Then her usual idea of herself changed to one in which at least two personalities were in her body. This model aligned with the cues that Antonio gave her about an imagined reality. It also eventually resonated with Anna, who might have thought that having other people inside could guide her to success.

So, is Anna one person or two? Most therapists would say she is one person—that she has only one "I," which they believe is the continuing essence of what makes her an individual. This is the way that most people think about our identities, one that aligns with most thoughts about mental health.

Yet Anna's own profound experience once told her that there were two of her. This was as real to her as was her previous belief that she, like virtually everyone else, was just one person.

As we will see, *both* of these beliefs are nothing more than models, conceptual understandings, of what a human being is. If we think of ourselves as having one, two, or even fifty different personalities, we are never necessarily right or wrong; it's just how we see ourselves at the time. We will also learn that none of these models are always useful. An important goal is therefore not to discover which model is correct. Instead, it is to understand both the nature of models and how one can best use each one in different situations.

It follows that the most important lesson from Anna, critical to understanding this book, is that literally every part of our realities is not only "made up," in a sense, but changes depending on our circumstances. Not knowing this, we might think we objectively perceive ourselves and the world. Anna, for instance, first thought it was obvious that she was just one person. However, her changing views about herself and her "minds" show that nothing people believe, including Anna's lifelong ideas about her own identity, is so simple.

> *Mirror's quiet gaze—*
> *Reflection shows the path,*
> *New dawn in old eyes.*

CHAPTER TWO

REFLECTIVE PRACTITIONERS
Insights and Transformations

Tailoring Treatment

Becoming a skilled therapist is no small feat. As a psychologist, I often struggled to understand the many ideas underlying psychological matters, each with its own unique opportunities and challenges. Also, the field is huge and includes many types of treatment. Some are effective and some are not. Most require years of training.

Furthermore, in therapy there is an intricate dance between science, religion, culture, and belief. Each makes counseling even more complicated. For example, one of my patients was a devout woman who rejected hypnosis. She was convinced it was against her spiritual beliefs, almost as if it were a type of blasphemy.

She is far from alone. That's why I entered her world by having accepted that her beliefs, different from my own, were nevertheless valuable. For instance, I considered weaving stories into her therapy, like the parable of the persistent widow.[1] This is a Christian Bible story about a woman who relentlessly hounds a judge for justice. He finally caves. This idea shows some religious people, like her, how persistence and determination can help move mountains.

Tailored treatments are essential at other times as well, but particularly so when working with marginalized minority groups, children,[2] the poor, and gender-diverse communities.[3] Picture them coming to you, as a therapist, for help with far more than run-of-the-mill stress, a challenge that often requires unique therapeutic approaches. With them, you might have the added responsibility of finding new ways to help them deal with social stigmas and hate crimes.

Undoubtedly, the therapist's education and expertise in tailoring therapies to each client are paramount. Yet therapists have their limitations.

Curiously, one therapist might easily navigate therapies focusing on patients' complex inner dialogues and yet falter grasping ideas about simple reward-based treatments. Moreover, therapists often naturally lean toward adopting therapeutic approaches that align with their own intellectual strengths and training, something that probably will not meet the needs of all patients.

Pivotal aspects of my work were recognizing my abilities and seeing where I fell short. Such insights, for me, were not one-time revelations but recurring realizations that led to exciting career-long explorations and struggles. These new understandings drove me to seek out more diverse perspectives, even from people with whom I first shared little in common. I soon found that the joy intrinsic to this broad learning was exhilarating.

We have seen that no one model of humans works well for all patients. Like Antonio, in my own counseling work, I've tried to select whichever models, and treatment methods, are likely to be most effective—for particular clients. When I was not likely to be able to help someone, I sent them to somebody else who could have more success. This mindset took into consideration that I too was limited, and that no single technique can help a psychologist skillfully manage the diverse complexities of individual patients.

And even with my monumental efforts to improve myself and customize treatments, I saw that everything I did, and every treatment model, had shortcomings.* Each one, even when useful, was incomplete and had built-in contradictions and weaknesses.

For example, a therapy that focused solely on changing observable behavior might have neglected the long-past causes of current emotional distress. On the flip side, therapies that strongly encourage patients to express emotions, while sometimes effective, can hurt other patients struggling with anger control.

These are multilayered issues that go beyond the realm of therapy. Choosing models wisely can be central to solving many of our everyday dilemmas. In both life and therapy, we are usually better off if we are adaptable, have wisdom, and know that no one approach holds all the answers.

Such challenges are not unlike struggles in many fields. Scientists, scholars, artists, and others often learn that the enormous complexity of the world forces them to rely on different types of math, physics, or observations.

---

* As we will see, this characteristic applies to *all* models, including those that I have created.

Models also evolve, often being modified or replaced with the acquisition of new knowledge. In the end, no model will always be useful or ever fully capture the world around us.

In this book I discuss how we use models to help us decode behavior. This does not imply, though, that I know people really make them in their heads. To do so would be to make conclusions about things nobody can see. But, talking as if people do make models in this way is, in fact, an often-useful way to imagine human thought.

Therapist Growth

The professionals whom I trust the most realize the limitations of their concepts and knowledge. They are refreshingly humble. Most know that no model is an absolute truth that will never need to change. They also know that we will always be partly guessing about what will best facilitate each client's healing.

They see how their own overconfidence can blind them to new perspectives. Such insight often translates into an ability to adapt and then shift from one approach or idea to another. Most are lifelong learners, never reluctant to question, and are people whose views evolve as new information becomes available. With acceptance of change, they inspire others to follow suit and innovate.

When I was a student, I encountered the example of Anna and Antonio. At first I thought it was a testament to the benefits of an innovative therapy.

But in retrospect, the most important lessons this case taught me, like it did Antonio, were about humility and our brains' amazing capacity for thinking about ourselves and the world in multiple, and potentially adaptive, ways.

Early Years of Training: The Fixer

For many good students, experiences such as having worked with people like Anna will forever influence how they relate to their future patients. To bring these changes to the forefront, and to focus budding therapists' attention on a critical issue, I recommend that the following question be asked of students each year of their training: What do you consider to be the most important job for yourself as a therapist?

If my professors had asked this of me, I could have better seen the evolution of my thoughts. Every answer might have been like rings in a tree trunk. Each, hopefully, would have shown a season of growth.

Early on, my answers would have been far different from what they would be today. As a young and naive student, I probably would have said, "I'm going to show people how to feel better." Of course, there is truth in this idea. However, this narrow perspective would have emphasized a gap between myself and my patients. It would have revealed that I saw myself essentially as a repairman working on someone who was broken. It would have been like Antonio would have felt if he thought Anna just needed his "expertise" to improve.

But this "I fix" mentality, if taken too far, erects a barrier. It is an approach that is not ideally respectful of patients' dignity. With experience, I came to see treatment somewhat more like explorers going on a journey together. This shift in my approach was helpful and better emphasized collaboration.

Antonio might have felt a similar change in perspective. I'm hoping that he did not just learn about Anna's complex problems but also about the limitations in, and promises of, his therapy. He could even have changed subtly near the end of his work with Anna and become less like a therapist following a textbook treatment. Ideally, his view of himself as the fixer on one side and the sick patient on the other would become blurred.* With this, *he too would have seen himself differently.*

Anna's story is not just an example of how malleable our brains are, but it also focuses on how fluid therapists' roles can be. It shows how our broader societal roles, and how even therapists' views of themselves, change. It is one tiny example of our far broader experiences in the world.

Reimagining the Mind

This book delves into common, sometimes helpful but often unhelpful, models of the idea of *mind*. It looks at how these ideas help us, harm us, or in many cases do both.

---

* Therapy is like a two-way street, a give-and-take between the therapist and the patient. If you're planning to be a therapist, go easy on yourself when you look back and cringe at your rookie moves. Don't think of your early work as having been only a string of mistakes. It was the time where you grew and refined your skills. You met new challenges. You got better at helping people heal. So, in the future, you can pat yourself on the back, knowing how far you've come.

I also explore our collective scientific and philosophical knowledge (and guesses) about what a human mind might be and how it works. I examine how the mind itself is sometimes thought of simply as the brain, and other times as something much larger.

Lastly, I offer strategies that could serve us well in the future—and that might prove more valuable than most of the models we currently favor.

Challenging these and other concepts will undermine many of our comforting ideas about human beings, including the following:

- *Our thoughts control our actions in clear, observable, self-evident ways.* Instead, many things we do are not supported by rational thought.
- *Each of us can observe our own thoughts.* In fact, our imagined observations of our own thoughts and feelings are illusory.
- *We have clear insight into ourselves.* Instead, many things that shape our behavior go unseen.
- *Thoughtful consideration precedes most or many of our actions.* Yet many behaviors could be better classified as automatic and unrelated to rational thought.
- *We have free will.* Understand instead that this belief is only a model and that our actions are determined by biology, our environment, and history.
- *A trance state is a stark departure from our "normal" state of awareness.* Instead of seeing a clear distinction between the two, we can view them either as the same thing or on a continuum. Doing this, we might visualize them as having varying degrees of differences from, and similarities to, each other.

By examining different concepts regarding the mind, we will come to a more realistic view of psychology and of our essence. This knowledge, in turn, holds promise to help us act more wisely and be more empathetic with others. Crucially, it might also help us to develop humility—a quality that nearly all of us can use.

In this exploration, I also show how some of our common models are difficult to substantiate. By showing their weaknesses, I'm not dismissing them—not at all. But we need to recognize that something—a model, for example—can offer insights, predict events around us, help us make useful decisions, and even be sensible, without being true. Thus, we see that a model, which is necessarily a simplification of reality, can still be worth embracing. As we will see, none of them always holds up to serious (or in some cases casual) scrutiny.

And so, despite their shortcomings, we know that each mental model might hold value—and each has the potential to tell us deeply important things about how we think and what it means to be human. Few people would now say, referring to the myths of ancient Greece, "They're literally true. There was an actual man named Narcissus who fell in love with his own reflection. And there was a real-life guy named Sisyphus who spent all his time pushing a rock up a hill." Nevertheless, we would be foolish to ignore what these myths—and a great many others—can teach us about ourselves. They can, even when literally false, hold profound and useful insights. In some people they trigger useful thoughts on topics ranging from morality to philosophy.

Seeing our models of the human mind as guiding ideas, instead of as accurate descriptions, can lead to richer knowledge of ourselves and the world. However, gaining understanding requires us to think much differently, and to have a willingness to experience some discomfort, about who and what we are.

Such a transformative change in our thinking is not easy, in part because our usual ways of thinking are grounded in comfort and practicality. Also, acknowledging that our beliefs should change to adapt to new knowledge can create insecurity. There can be a reassuring warmth in believing that we have ourselves and the world figured out.

But there are other benefits to having a more authentic understanding about ourselves. Doing so, among other things, can open us to new perspectives, help us enjoy the diversity and richness of our existence, and ignite our curiosity. And finally, seeing that our models are not absolute truths can give us a larger and more helpful perspective for learning about, and coping with, the world.

In this book, we will discover that we are far different from what we usually imagine ourselves to be. And so, I invite you to leave your familiar and comfortable views aside and join me on a revealing journey to explore our nature.

Growing thoughts take root,
In wisdom's soil, they mature.

CHAPTER THREE

MODELS GUIDE AND MISLEAD

W e generally create models that help us navigate our current circumstances. For example, if we're floating in outer space, we don't need to think of the world in terms of up and down, because they have no relevance. But once we return to Earth, with ground below our feet, that way of thinking becomes all-important.

Such models of the physical world sometimes begin at the subatomic level. We have models that describe how particles form atoms, how those atoms bond to create molecules, and how those molecules combine to produce compounds. Others help us create new chemicals and learn their effects on humans. With this knowledge, scientists have developed many lifesaving drugs.

Psychological Models of Behavior

We have seen that psychologists employ models too. Some describe people as flesh-and-blood machines, with various parts influencing each other in complex patterns. In some of these approaches, the human "machine" behaves according to the laws of physics.

Freud famously observed that even young children forge models of themselves and their environments based on their experiences. For example, children sometimes conclude, based on limited experience, that all men are dangerous, or that all mothers are loving but devious. Freud also noted that children sometimes continued to believe their early ideas well into adulthood, despite contradictory later experiences.

Other psychologists have devised models that emphasize how events in the outside world influence human behavior. Still others focus on brain waves and heart rates during hypnosis.

Psychologists, scientists, and philosophers have created many models that describe other unseen mental operations. Consider the following example from biologist Richard Dawkins:

> When a man throws a ball high in the air and catches it again, he behaves as if he has solved a set of differential equations in predicting the trajectory of the ball. He may neither know nor care what a differential equation is, but this does not affect his skill with the ball. At some subconscious level, something functionally equivalent to mathematical calculations is going on.[1]

Here Dawkins invokes a model that draws a functional parallel between what a mathematician or a computer does and the activities in the brain of a person catching a ball.

Scientifically sound models are sometimes the most helpful—but not always. And, as we will see, *every* psychological model has turned out to be limited, if not also partly wrong.

Models and Social Dynamics

We sometimes choose to use a model even when we *know* it is wrong—because it helps us navigate our culture, community, workplace, or profession.

Indeed, most of us grapple with the pressure to follow approved culturally specific models, even when they directly conflict with science. For example, some communities, such as some of those in the Caribbean, encourage people to accept and follow practices of voodoo.

In certain situations, using accurate but less accepted models can be dangerous. In 16th-century Europe, the predominant teachings explained that the Earth was the universe's center, placing humans above nature. Copernicus (and others) realized that this was incorrect and that a model of the Earth orbiting the sun was accurate. Against all the then-current thought, his theory suggested that the Earth, and humankind, were not superior to nature but part of it. Nevertheless, back then, if you were a prominent scientist and didn't want to be persecuted, you were better off believing (or pretending to believe) in the official, pre-Copernican concept.

Similarly, in our own lives, our beliefs about what is in our best interests can influence our models. For example, in some groups and contexts, we may be lauded for sharing a profound spiritual experience; in others, we may be laughed at, worried about, or shunned.

Anna and her therapist described her experience and actions as "hypnotic"—a label that offered her physical and psychological benefits. Conversely, a faith healer could easily attribute what happened to Anna to God's power to heal. As it could enrich their lives through a sense of shared fulfillment, such a narrative could thus reward Anna and the faith healer.

Scientifically unverified models of altered states, disease, and the power to cure have been used for millennia—occasionally with good results. For example, for many centuries, the idea of demonic possession was used to explain physical illness—and to help certain community figures gain and hold power.[2] (Notice the parallels between these "demons" and Anna's imagined childhood tormentors.) Exorcism, the "cure" for possession, promised a variety of medical benefits, but only to members of various devout groups who accepted that concept.

Exorcism didn't just benefit people who were thought to be possessed. It was also an effective recruitment tool for spiritual followers, a way to label enemies as demonically possessed, and an avenue toward attention and fame for the exorcists.

And, as with public faith-healing demonstrations today, public exorcisms were a useful way to prove the power of faith—and the power of the exorcist.

Those in power often spread models that serve their interests. For example, it's common for them to encourage the belief that strong social hierarchies are essential to a functioning society. This viewpoint says that just a few people should legitimately have more power than others because of their wealth, gender, religion, genes, or lineage.

This helps those in power both obviously and subtly. For example, Matthew J. Hornsey and Kelly S. Fielding[3] revealed that people who believe in strong social hierarchies are likely to dismiss science and be skeptical about climate change. Possibly this is because this skepticism does not challenge the interests of some people, such as powerful oil company executives, whom some less affluent citizens think deserve to be in control.

Even Anna must have struggled with her own preconceptions, as well as with societal pressures. After her positive, and in some ways transformative, experience with hypnosis, she was at a crossroads. She could revert to her old models, which were simple and based on false assumptions about both who she was and the cause of her problems. Or she could instead embrace the more complex insights she had discovered during therapy. It was not a straightforward choice. Social stigmas about multiple personalities, for instance, which are themselves models, might have influenced her decision.

We know that such tugs-of-war between social expectations and personal beliefs aren't unique to Anna. They are something we all encounter. Outside pressures heavily influence our models about many things. These range from politics to health care and abortion. Just as Anna likely had to wrestle about whether to embrace her newfound insights, everyone else must navigate the tension between what we think is true about ourselves and who others expect us to be. Anna's story, therefore, is a small-scale version of a much larger, and often messy, struggle between individual thoughts and collective ideology. Our growth always takes place in a much wider, and important, social context. So, as we work for greater insight about ourselves and the world we live in, remember that our models are never formed in isolation.

Model Transfer

In the mid-1950s, well-known psychologist B. F. Skinner[4] noted that when we adopt a model, that concept partly defines our environment. Then, when we share this model with others, we in turn define *their* environments. This phenomenon is what I call *model transfer*.

I am talking about, of course, a two-way relationship between individuals and their communities. We have seen that when Anna discovered "Sarah," it wasn't just a change in her understanding of herself. Anna also became a type of ambassador, showing Antonio that hypnosis can be a tool in therapy, not just for herself, but also for others.

Model transfer can be a selfishly beneficial tactic. In social and interpersonal contexts, it can be used to influence, manipulate, exploit, or control other people. We can use model transfer as a tool to help others—or as a weapon to harm them.

If I start a long conversation with you about the importance of donating money to the Nature Conservancy—and, after our discussion, we each mail a check to that organization—I've done much more than raise funds. I've also gotten you, at least temporarily, to adopt this model: *The conservation of nature is important and wise—so we should support it with cash donations.*

Similarly, if a political candidate promises you that they and only they can bolster our nation's economy, this is not a mere transaction in which they win your vote and your life remains unchanged. If you believe their promise, you have also adopted the model that they are uniquely skilled and powerful—at least when it comes to economics.

Model transfer—and the ability to clearly convey an idea to others—can be powerful. Over millennia, people who could most effectively implant their preferred models into other human beings' heads got to survive, thrive, have kids, raise many of those kids to adulthood, and teach *them* the techniques of model transfer.*

Note, of course, that the effectiveness of model transfer might not be contingent on the concept's validity. A useful model could be highly accurate, pure fantasy, or something in between.

Steve Jobs: A Master of Model Transfer

Steve Jobs, the cofounder of Apple, scored major success partly because he used the concept of model transfer so well. The term "reality distortion field," a now popular and catchy phrase that comes from the *Star Trek* TV series, describes his use of this idea. On *Star Trek*, the phrase painted a picture of a make-believe distortion field that changed reality. Here, I use it to describe Jobs's charismatic and seemingly magical ability to get others to buy into his vision. With this skill, he could bend the perceptions of those around him to align with his own.

Even Bill Gates joked that Steve Jobs used reality distortion fields to "cast spells" on people. He claimed to be immune to this influence, explaining that, "because I'm a minor wizard, the spells don't work on me."[5]

Picture Steve taking the stage wearing his jeans and turtleneck. He was selling a vision of how the world should be, not just a gadget. Instead, his presentations were invitations to adopt his sometimes distorted, yet appealing, versions of reality.

With his incredible skills, he introduced game changers such as the iPhone. He got his technical teams to engineer what was thought to be impossible, and for people to stand in line to buy his products. He

---

* The destructive power of model transfer is on clear display today. In one widespread model transfer, politicians convince voters that vast hidden networks of people are secretly plotting against them (Hornsey and Fielding, 2017). In another, conspiracy theorists insist that scientists seek to harm us and that COVID-19 vaccines kill rather than protect us. This latter contention encourages people to not only avoid inoculating their children but to discount science in general (Lewandowsky et al., 2015; Lewandowsky et al., 2013).

transferred his models and other beliefs to his entire industry, even reshaping global culture.

And so, while Antonio's use of model transfer helped Anna, we see that Steve Jobs's models took things to a whole other level, having a massive impact on society. He helped us see new potential in technology. This was a tactic that pushed us to reimagine our lives.

Getting the hang of model transfer can help us be savvy in navigating the world. Being aware, for example, of the influence a charismatic figure like Steve Jobs can have on us, good or bad, helps us adopt both curiosity and skepticism.

Useful but Scientifically Weak Medical Models through the Ages

The pioneering Greek physician Hippocrates lived over two thousand years ago. Before his time, the usual Western models of disease portrayed demons as the cause of physical illness and often prescribed prayer or sacrificial offerings as treatment.[6] People from some other ancient cultures took an even more gruesome approach, cutting holes in patients' skulls to release evil spirits.

The people of Hippocrates's time who followed most Western models related to demons were often certain that those concepts were correct, and they sometimes tortured those who felt otherwise. These spiritually based models were nevertheless useful—at least sometimes for maintaining power structures and keeping people in line, though probably not for curing disease.

Hippocrates bravely challenged these ideas with a new approach to medicine, which stressed observation and physical cures. He helped create the first scientifically based medical approach to healing. For instance, he was the first person to describe epilepsy as a brain disease rather than as demonic possession.

As the years passed and scientific explanations for disease gradually replaced those of demons, it became less important for sick people to show characteristics that suggested possession. The commonly accepted models, and the local power structures, therefore permitted these people to simply be ill. This was a monumental shift, as the sick no longer had to bear the weight of imagined demons.* It also laid the groundwork for medical science.

---

* There are some notable exceptions. In voodoo today, practitioners are often "ridden" (i.e., temporarily possessed) by spirit-like *loa* and dramatically change their behavior for minutes or

Despite his great advances in thinking about illness, Hippocrates was not infallible. He thought that illness occurred largely because of an imbalance of four bodily fluids, or "humors": yellow bile, black bile, phlegm, and blood. He thought that excess black bile caused sadness, lots of blood caused cheerfulness, too much yellow bile made people crabby, and extra phlegm resulted in apathy. He often treated his patients by trying to adjust the imbalance of these humors with such methods as diet, lifestyle changes, vomiting, bloodletting, and herbs.

Even though Hippocrates's ideas were sometimes misguided, his followers stuck to them for many centuries. In 1776, Baron Gerard van Swieten[7] was still writing about bile and blood in Hippocratic terms. (To cure some mental problems, van Swieten recommended throwing people into the sea, or giving them mercury to change their alleged imbalance of unhealthy bile.) On a very few occasions, some cures based on Hippocratic models appear to have worked. Perhaps these were early examples of the power of placebos.

Other seemingly scientific models that were once widely accepted were essentially affirmations of their originators' skills—and/or the dehumanization of certain groups. Phrenology, the study of detecting individual traits and abilities by measuring people's head shapes and facial characteristics, was popular from the mid-1800s through the early 20th century. Areas of brain tissue and the skull, each of which varies in size from one person to the next, were thought to reflect greater or lesser degrees of brain development. People who had the most favorable phrenology ratings had large, broad skulls, high foreheads, eyebrows set wide apart, and large eyes. Dr. Francis Gall, one of phrenology's founders, happened to have these traits himself. (What luck!) Many phrenologists believed that non-Caucasians routinely had some of the most undesirable head and facial features.

So, diving into the world of phrenology gives us a bit of a reality check. Gall seemed like he was all about science, but his model had ugly undertones. It not only propped up his own theories but also gave a scientific veneer to some deeply ingrained racial biases. That is a lesson to remember. Even when something is dressed up in fancy scientific language, we've got

---

hours. While being ridden, some people dance or convulse; some adopt different personalities. The Japanese religion of Shinto includes a belief in demons, and Shinto priests today routinely conduct ceremonies similar to exorcisms.

to be careful. Sometimes what looks like objective research can instead be a sneaky way to push an unsavory agenda.*

We have, of course, seen such misuse of science in modern times. Both conservative and liberal groups in the United States have manipulated, or ignored, data to support their own agendas during the COVID-19 pandemic. One example is when conservatives recommended the antimalarial drug hydroxychloroquine for the treatment, and prevention, of COVID-19. To do this, they usually relied on preliminary studies that had little peer review. Later solid research found this medication to be useless, and often deadly, for this purpose.[8]

On the flip side, liberals sometimes overstated the risks of COVID-19 for certain younger and healthier groups. For instance, they were known to gloss over data showing that the elderly and those with preexisting respiratory conditions were most susceptible to the effects of COVID-19. Misusing data once more led to confusion and flawed public policies. In both cases, cherry-picking information or using bad data that only seemed scientific in origin led to confusion and conflict.

Social Pressures and Rational Thinking

Western cultures typically uphold a model suggesting that most people are rational most of the time. This belief also specifies what constitutes "rational" and what does not.

For instance, if we routinely believe we are two (or more) different personalities rather than one, others often think of us as irrational, deluded, foolish, and/or mentally unstable. Yet, if this same belief emerges during a hypnotic trance, those same folks might well perceive that belief as therapeutic.

Superstitions are another common, often accepted, departure from reality. Think of it: People have been to the moon, but we still knock on wood to avoid "jinxes." Remember movies, for example, where superstitious people are portrayed as charming. Maybe they look for a four-leaf clover and their luck returns.

In the West, the cultural importance of rationality likely has some roots in the Enlightenment era. Think of it as having been a Woodstock for

---

* Sadly, developments in artificial intelligence have given rise to researchers who once again claim to determine tendencies toward criminality based on the shape of people's faces. And so this struggle continues.

intellectuals. It occurred in the 17th to 18th centuries.[9] This "Age of Reason" was both about breaking free of superstition and traditional ways of thinking and instead embracing science. It was a time of productive rebellion. And so, rationality soon became a cornerstone of Western culture. It was the new gold standard. Even today, to fit in, we need to support this model of appearing as reasoned beings.

In modern times, we can't have others thinking that we are mentally unstable. So, our informal Western model of ourselves is often something like this: *I'm normally rational, except for therapeutic purposes when I temporarily change under the influence of a hypnotist.* Or, in other contexts, *I'm normally rational, except when I speak with spirits, but that's okay with my religion.*

Note what's going on here. We justify feeling a breeze in a windowless office, suddenly speaking in an unknown language, or getting luck from a good luck charm by carving out excusable exceptions to rationality. The causes of these exceptions—superstition, hypnosis, therapeutic technique—vary widely, but they can all serve the same purpose. They can be seen as justified deviations from our presumed "regular" state of reason. They are, in effect, "rationality escape hatches." Such approved exceptions help us support impressions that people are generally logical and reasoned beings. We use them to maintain a collective story, or illusion, that we live in stable, rational societies.

This all suggests that the practice of sanctioned departures from rationality is likely more common than we first realize. And that's not necessarily a bad thing, because it's part of being human. Sometimes our immediate emotional needs want the fast foods of love, sympathy, and validation of even incorrect ideas. We do not want to wait around for the fancy fare of a five-star restaurant. We see, then, that it's not as if people always abandon the appearance of rationality. But we seem to pick and choose the times when we want it. Knowing this, we see that our lives are not navigated by reason alone.

The Porosity of Self-Perception: Human Biology and Models

We humans are biologically wired to use our sense organs to categorize the world. These categories, or models, create divisions such as tasty versus unpalatable, beautiful versus ugly, honest versus deceptive, and us versus them.

Our senses of sight, hearing, smell, taste, touch, and balance facilitate such natural divisions. This shows how we base parts of many of our ideas not just on what has practical benefits but on our biological signals, abilities, and predispositions.

For example, in judging people on the beautiful-to-ugly continuum, most of us primarily use our sense of sight. But if we could detect the electrical impulses coming from other people, as a platypus can, we might instead (or in addition) use those fields to judge others' attractiveness—and, perhaps, their health and personalities.

Our biology also helps us model one of our most basic divisions: ourselves versus the outside world. We tend to see ourselves and the rest of the universe as fundamentally separate things. Indeed, we experience this distinction as both innate and obvious. But in fact it is neither.

In a study on subjects with normal perception,[10] each person watched a mannequin's rubber hand on the desk in front of them. The person's own hand was hidden by a screen. At the same time, the experimenter briefly simultaneously touched both the mannequin's hand and the subject's hidden hand with a probe. This was repeated every few seconds. After a few minutes, about half of the subjects experienced the rubber hand to be part of their own bodies. Their brains created a model of reality in which they incorporated the rubber hand into their concepts of themselves.

This is compelling evidence that our perception of a "self" can change and depends at least partly on sensory input. Our identity, therefore, is intricately woven into the changing sensations we encounter.

But there's more. This rubber hand phenomenon, called *illusory ownership*, didn't just happen when subjects' hands were touched. For some of the subjects, their brains took an even bigger leap. They didn't need to be touched for the illusion to work. In fact, when they merely *observed* the mannequin's hand in about the same position as their own, they took ownership of it as if it were their own flesh and blood. They were able to bypass sensory input to manipulate their sense of self.

And so, our changing perceptions of ourselves can also occur just by inference and observation, rather than by touch. This should amaze us, seeing more about the complexity, and power, of human thought. It reveals that our sense of self is more porous than we realize, and it depends a lot on what we see, and our beliefs.

You can easily conduct a similar experiment in your own home.[11] All it requires is you, two chairs, and two other people, one of whom should be roughly the same height as you.

Place the two chairs about two feet apart, one behind the other, each facing the same way. Sit on the back chair and have the person who is about as tall as you sit on the other with their back toward you. Have the third person stand next to you.

Then close your eyes. Slowly reach forward around the other sitting person's head and lightly touch the tip of their nose for a second or two. Then pull back your hand briefly. Rhythmically repeat this process over and over, lightly touching their nose, then pulling back, then lightly touching their nose again. Keep your eyes closed the entire time.

As you do this, each time you touch the other person's nose, the third person should lightly touch *your* nose. When you remove your hand, the third person should remove theirs. As much as possible, their movements should mimic yours.

After a few minutes, when you touch the other person's nose, you might have the experience that you are touching *your own* nose, which will now seem two feet long. No wonder it's called the Pinocchio illusion.

Hobbyists who fly drones while wearing "first-person view" goggles describe a similar change. These people see things from the perspective of the drone as it flies through trees and around buildings. Many of these hobbyists feel as if *their own body* is flying; some experience a sense of somehow *being* the drone.*

A multitude of other situations can call our usual perception of self into question. Many people have experienced so-called "out-of-body" episodes, in which their "souls" or "centers of awareness" seem to float above their physical bodies or move through walls or doors.

As obvious and self-evident as the self-versus-other distinction may be, it is just a model rather than an unchanging reality. Where you sense that your body starts and where it ends can change depending on many things.

---

* Certainly, merging ideas of the self with technology is controversial and raises many ethical concerns.

How Evolution Influences Our Models

One widely held theory is that the people whose thinking skills allowed them to devise and use the best models have passed those skills—and those ideas—down to their descendants. This theory suggests that early humans lacking the cognitive and perceptual hardware to make the most useful distinctions gradually left the gene pool.

While the fossil record does not give us details about the evolution of thinking skills, we do know that, over millennia, the human brain has evolved. For example, the skull sizes of Neanderthals suggest that they might have had brains about the size of ours today—and perhaps some skills superior to ours. At the very least, they appear to have been capable of abstract thought and language, since evidence suggests strongly that they followed rituals when they buried their dead. But were they *as* skilled as us at language? At forming abstract ideas? At using models? We don't know.

As our brains became increasingly intricate and specialized, they have enabled such things as communication using speech, abstract ideas, and the development of countless models. Like all adaptive traits, those thinking skills that promoted human reproduction, and therefore the transmission of an individual's genes, would have been the ones most likely to become established within our species.

Freud's New Blueprint

In the early 20th century, psychoanalyst Sigmund Freud[12] introduced a groundbreaking model of the human psyche, which included the *id*, the *ego*, and the *superego*. While these concepts were innovative, in some ways they paralleled Hippocrates's four bodily fluids. Like those fluids, the id, ego, and superego were all used to explain many illnesses—and they supposedly interacted with each other in ways that nobody could directly observe.

According to Freud's early model, the id follows instinctual urges, such as those for food, physical comfort, and sex. The id is utterly selfish. The superego often stifles the id's urges with what many have described as civilization's "conscience." The id and superego are in almost constant battle. The third entity, the ego, helps people cope with practical matters of daily living and helps the id and superego get along with each other. In short, Freud attributed humanlike qualities to each of these proposed internal entities.

In Freud's model, unresolved conflicts among these three entities cause psychological suffering, which can lead to mental illness. His techniques of psychoanalysis were designed to help these three parts of the psyche resolve their differences.

According to this early model, some of our true motives, such as many of our wishes and desires, supposedly remain buried in an unconscious mind, like a secret note we might find under our mattress.

When Freud first presented his model, it was wildly new and revolutionary. It fundamentally altered the way people thought about human consciousness, and it ultimately enriched our thinking about human nature.

Freud's model creatively attempted to explain what goes on inside our heads in terms of what he saw every day: interactions between people. Although he never directly observed battles between an id and a superego, he inferred them.

It's fascinating how Freud gave human traits to psychological concepts. In some ways he turned unseen portions of human experiences into characters from a drama. By doing this, he cleverly made studying psychology more relatable.

But let's remember something important about Freud: His own world was a melting pot of shifting social norms and sexual hang-ups. He was living in a Victorian society where everyone had a ton of rules, especially about what you could and couldn't do based on your gender or sexual orientation. This atmosphere of rigid decorum probably seeped into Freud's theories about sexual repression and the conflicting facets of human nature.

Freud's theories therefore weren't just random musings. They were shaped by the world in which he lived. They were influenced by the societal upheavals, sexual repression, and daily struggles of the people around him. We see then that science can be a mirror, reflecting the unique challenges and beliefs of the era.

In no time, Freud's model caught on. In fact, Freudian psychoanalysis and its offshoots dominated the practice of psychotherapy for most of the next half century. His work not only helped catalyze the entire field of modern psychotherapy but spurred a great deal of important research and debate. His influence remains, with many, but not all, of his theories still being endorsed by professionals.

A Conventional Commonsense Model

"Commonsense" models are not built upon careful scientific investigation. For example, one theory of how the human mind works goes like this:

> When people do things for inexplicable reasons, their actions are often driven by desires hidden in the unconscious mind. To help them better understand why they do things they can't explain, they sometimes benefit from delving into the normally hidden and inaccessible corners of their minds. This introspection can help them discover the nature of these concealed forces. Such exploration can be therapeutic, especially when it reveals forgotten traumatic events that cause present distress. Hypnosis often facilitates such introspective journeys by enabling people to confront these destructive unconscious memories.

This theory merges some aspects of Freud's model of the psyche—focused on human interactions—and our own interactions with *objects*. Here, it presents minds as having hidden places for desires, somewhat akin to cubbyholes in a desk. Some hope that, with close inspection, a therapist will help uncover previously unknown issues hidden within these mental alcoves.

Decades of experiments suggest that, while some elements of this model appear to be very useful, others are questionable, and some are largely useless. Nevertheless, many in the Western world still often successfully follow Freud's groundbreaking teachings.

Some theorists do this when they explain why people deny responsibility for their actions. They can infer that people deflect blame and responsibility when they are troubled by their own motives. For example, if I am not upset that I have hurt someone, they might conclude that I have unconsciously blamed my hurtful actions on someone else. Such false explanations can be thought of as keeping knowledge of my original plans away from my awareness, consequently preventing psychological problems. As a result, they might believe that my true motives—my hidden wishes and desires— remain buried in my unconscious.

Therapist and Social Media's Roles in Self-Modeling

People tend to base their models of themselves partly on what they are told by mental health professionals, and partly on what they encounter in the media.

Let's first look at social media, which are like digital town squares. All of us gather there for the latest news. But here's the thing: These platforms often cater to specific groups. Almost regardless of your beliefs, there's a space on the internet made just for you.

These spaces can be lifesavers for marginalized people. That's because when you discover your own online neighborhood, you'll most likely experience feelings of belonging and validation.

But when you're always hearing the same ideas and opinions, you're essentially living in an echo chamber.[13] Think of it as a hall of mirrors. Instead of seeing yourself from different angles, you only see one reflection, and that's your own. Because of this, you could start thinking that your perspective is the only game in town.

And what's more, these echo chambers often give voice to strange or even dangerous beliefs. Whether it's conspiracy theories or shady "miracle cures," these platforms can empower people pushing fantasies. So, while they can help like-minded people band together, they sometimes also reinforce distorted worldviews.

It's also understandable that we create some models of ourselves based partly on what mental health professionals tell us. Who has the time, energy, or information to create from scratch their own model about the human psyche, human relationships, and everything else? We naturally borrow from what we hear from professionals.

Let's return to Anna for a moment. Her experience during hypnosis was guided by the model that Antonio presented to her. It might also have been shaped by social pressure to lose weight—and, perhaps, by her desire to gain Antonio's approval.

In modern times, when people regularly see themselves as having more than one personality, they are often diagnosed as having dissociative identity disorder (DID), formerly called multiple personality disorder. Originally, DID was thought to have been the result of trauma[14]—and, often, of sexual abuse, which might be forgotten or buried. (Many therapists wrongly told clients with DID that the disorder's cure usually depended on their ability to resolve forgotten sexual trauma.) More recently, however, this idea has evolved, as DID is now viewed as largely a product of what people learn.

In the late 20th century, the United States witnessed a DID epidemic after Americans were widely introduced to the disorder—and then coached in it—by the media. In particular, two bestselling books, *The Three Faces of*

Eve[15] and *Sybil*,[16] taught us how to emulate DID by changing our voices, handwriting, food preferences, and moral standards. During this same period, in areas of the world without such examples, there were few or no cases of DID.

The number of personalities involved in DID cases in the United States has also changed over time. In the year 1900, people with "multiple personalities" rarely had more than two selves. As the decades passed and people had more exposure to media portrayals of this problem, however, patients often appeared to split into three, four, a dozen, or even hundreds of different "selves."[17]

Historical Echo: Charcot's Model of Hysteria

Something similar occurred in the 19th century. French neurologist Jean-Martin Charcot was a physician who championed a model of a mental illness called *hysteria*. This illness was thought to have included such symptoms as depravity, hallucinations, seizures, mood problems, and paralysis. This malady was believed to be especially common among women and was sometimes thought to be caused by neurological or psychological problems or, bizarrely, by a "wandering uterus."* In the hospital where Charcot worked, female patients had many opportunities to observe how they should act to promote his beliefs. They learned to become passive and produce seizure-like behaviors, in some cases on cue.[18] In a common and eye-catching demonstration, women would fall backward and seem to lose all awareness of their surroundings. Then their arms and legs would shake. Then, after some relaxation, their backs would arch, and they would assume postures that suggested crucifixion. Sometimes they would start praying. In other cases, they would become delirious.

Some of these patients consciously faked their symptoms—and later admitted to doing so. But others experienced their symptoms as genuine.

Charcot became a celebrity for his public demonstrations and "cures," which involved electricity, magnetism, and hypnosis. His patients desired to be in his good graces and, within the confines of his medical facility, were permitted (and encouraged) to act in ways that they could not in the outside

---

* Notice the parallels here with Hippocrates's bodily humors. Although the human uterus is of course real, its identified behavior—i.e., its ostensible tendency to wander—is imaginary.

world.[19] Accordingly, the model of hysteria was closely embraced by both Dr. Charcot and his patients.

Natural Selection and Juggling Competing Models

Many of the models we've explored are at odds with one other. Over time, some of them overshadow others, in what resembles an evolutionary process.

Often, however, many of us easily move between conflicting theories—and switch from one model to another as our situations change. Our knack for juggling different models truly shines in social contexts. It's as if we're born socialites and have thrived in groups partly because we have mastered complex, and often changing, rules of social etiquette. But our success is not just because we've learned to follow these expectations. It also depends on our ability to predict, and even change, others' behavior. Our talents therefore stretch beyond comprehending and following models. They also include interacting with, and responding to, the hints we pick up in social settings. For instance, let's look at how we read people. While one model helps us see someone's gestures—perhaps a gift of a bunch of flowers—as sweet, another model can suggest motives of hidden self-interest. This might happen, for instance, if we were raised in families where gifts tended to be enticements for future favors.

This skill in deftly switching between often diverging possibilities has not only given us an edge personally; it has also strengthened our communities. This both complex and detailed understanding of our neighbors' perspectives has likely both fostered teamwork and allowed the amicable resolution of conflicts.

Another example of our versatility is that when we're in science class on Friday, we are invited to accept that everything—every rock, atom, and solar system—is always changing. But on Sunday morning, when we're in church, we're shown the idea of an eternal soul. What a contrast!

There is another good reason to believe that this ability to nimbly shift from one model to another gives us special powers. The environments in which humans have lived have changed radically over many millennia. Visualize being a cave dweller with the Ice Age approaching. Gradually, your old ways of finding food and keeping warm don't cut it anymore. Given this, people with the most elastic thinking skills would likely have had the best shot at surviving.[20] It's this mental flexibility that probably gave us all a nice evolutionary edge.

Indeed, the ability to juggle many ever-shifting and contradictory models may have been essential to human survival. Natural selection and evolution have wired us for prioritizing success—i.e., survival and the opportunity to pass on our genes—and less for being sticklers for truth.

Oversimplified Models

One of the major hiccups with models is oversimplification. Leaning too much on any one easy-to-learn concept, which is more likely when we're only paying attention to one perspective, leads us down this dangerous path.

Our love for keeping things simple probably has deep roots in our evolutionary past. To cope with a complex world and to survive, our ancestors had to quickly spot patterns, rather than get bogged down in the details. However, this knack for recognizing patterns, while often handy, can make us gloss over the finer details. Such a broad-brush approach, a kind of shortcut to learning, can make us skip details that really matter.

Here's where a psychological phenomenon called *confirmation bias* comes into play. It fuels our bad habit of oversimplification and acts like a magnifying glass that only focuses on what confirms our earlier beliefs. This sneaky, and common, bias shows its face when we are attached to a model and then only accept other information that supports those beliefs. It doesn't stop there, as we can also easily dismiss or misunderstand information that conflicts with our comfortable viewpoints.

Such bias not only cements our simplified, often off-the-mark beliefs. It can create barriers against letting in fresh perspectives. This keeps people from realizing new insights and breeds mental stagnation.

There are extra hazards when models are too well accepted in our communities. The fact that all, or most, people accept them can feed into complacency. It dims the chances that people with different types of expertise will contribute new ideas. The resulting tunnel vision can prevent "Aha!" moments, the heartbeat of innovation, that come from diverse and complex perspectives.

Even the very act of crafting models can lead us down the wrong path. By their nature, models encourage us to learn about our world by breaking it into simple little pieces. This way of thinking subtly sneaks our common-sense biases into models, which often rely on imagining things as separate from each other. Consequently, it is likely to stop us from seeing the beauty of the universe as a coherent whole, in the manner of some Eastern thinkers. But more on that later.

Simple Models Obscure a Dynamic World

Clinging to conceptual models, especially simple ones, can lead us astray in yet another way, this time by painting the world as a straightforward and somewhat unchanging place. *Many, for instance, seem to crave the simplicity of following one philosophy, political ideology, stereotype, religious text, or scientific theory.* This happens partly because models have a knack for condensing reality into easy-to-understand formulas. They are like mental shortcuts. Consider the way, for instance, that you think about the sun "coming up" every morning. This is an easy way to not only explain an astonishingly complex event, but also feed our brains' cravings for simplicity.

Yet the world, of course, is anything but simple. It's dynamic and always on the move. But we are not always fans of this unpleasant fact. Maybe this is because accepting this would mean that we would have to constantly go through the exhausting task of changing our beliefs.

Our cultures also do a great job of pushing models that suggest the world is both simple and stable. But of course, nothing truly stays the same. Ever. We grow, we die, continents move, countries fall, and species become extinct. Time-honored laws, traditions, and beliefs, however, promote the idea of permanence. Certainly, they can put a damper on seeing the world as it is, which is anything but stable and predictable. They further cement resistance to change, making us and our communities both stubborn and less adaptable. And by buying into such simplistic thinking, we risk selecting only information that supports our already-comfortable beliefs.

Therefore, although models give us comforting and stable frameworks to grasp, they can also deceive. They mask the dynamic nature of our lives and lure us into oversimplified views of reality. In this way, models are like friends who always tell you what you want to hear. They seductively invite us into a make-believe land of simplicity and stability.

Seeing things as stable also gives you a nice warm sense of permanent identity, perhaps as someone who is smart, determined, or a chocoholic for life. This perspective creates an illusion of control, but instead it is like navigating unexplored and often-changing terrain with an old map.

Flexible and Broad Models as Antidotes to Fear-Based Control

As we've seen, our personal lenses distort our perceptions of what is truly an ever-changing universe, one that holds many possible avenues for improvement. Sadly, some fall into the trap of seeing the world as set in stone, with only a few select leaders having the keys necessary for change. By doing so, they don't realize that they are making themselves into targets for manipulation. That's because manipulators can exploit their inflexibility and narrowness by promoting alarmist stories both to instill a sense of urgency and to enforce conformity. In effect, they've created a scary world for which only they hold the map. This is like living in a dream world that is constantly swayed by anxieties.

But we don't have to let anyone pull our strings. Instead, when we think of the world as ever-changing, with alternative perspectives to consider, we gain power. To achieve this, we draw on our own evolving ideas, a lifetime of experiences, flexibility, and a willingness to consider new solutions.

In an Eastern sense, we would fittingly see the world as impermanent, subjective, and full of possibilities. We would have embraced a mindset where we question everyone, even those who rule by fear. Gradually, perhaps through much struggle and reflection, access to better information, as well as collaborative efforts with others, we can learn to discern whether we are being shown reality or just spin.

We therefore can become free to make our own maps. In this light, frightening and static images lose their power. We learn to see through the scary illusions that others, in their attempts to control us, create.

Changing Models Can Be a Tall Order

Grasping the concept of gender can be fascinating: It's a journey that's both complex and personal. It's also an area where people often agonize trying to change their beliefs. In traditional Western views, people are generally cast according to a biological model. In it, we are either male or female. It's assumed that you are a male if you have a penis and a female if you have a vagina. We see cues reinforcing this simple concept everywhere, ranging from the men's and women's sections of clothing stores to restroom signs and the pronouns we use.

One problem in changing this, and many other deeply ingrained models, is that it's so easy to only embrace comfortable ideas. They tend to stick like coffee stains on a white shirt. But as we've learned more about how we helped create ourselves, it has become clear to science that this traditional view about gender was too narrow.

With this revelation, we saw that our self-concept—how we thought of ourselves—doesn't solely hinge on anatomy. While it does play a role, anatomy doesn't fully explain how people understand and express their identities, either as male, female, or many other ways. Culture has a say, too. Some societies, for instance, recognize more than two genders, showing how gender is a story that can be written differently across the globe. Realizing this led us toward more contemporary and fresh insights.

The American Psychological Association offers a more updated view of gender. It sees it as the "*roles and behaviors* society expects based on one's biological sex." In contrast, gender *identity* is more of an internal feeling, not just tied to biology. Our identity can be our innermost sense of being male, female, genderqueer, or an entire range of other identities. It's "a deeply felt, inherent sense of being a boy, a man, or male; a girl, a woman, or female; or a nonbinary gender."[21]

In much the same way that Anna changed her own storyline, questioning the biology-only view of gender can bring clarity and more authentic ideas of who we are. Gender is not just biology, but it's about our entire being and experience.

It's important to realize that when people wonder, "Why does that person identify differently than their biological sex?" they are not asking the right question. They are reducing gender to only biology. It's like calling someone with a naturally stern appearance deluded for feeling kindhearted. It doesn't seem fair, does it?

Creative Disruption: How the Arts Remodel Society's Models

So how else might we fix problems stemming from models? Well, supporting the arts can help. Authors and other artists have long been at the forefront of new social norms, presenting novel ideas and encouraging change. Some of these even inspire us to think differently.

Take, for instance, Frida Kahlo's self-portraits. They often blended masculine and feminine traits, defying traditional ideas of femininity. And then we have author Harper Lee, who challenged racial stereotypes and showed some of the pains of racism.[22]

More recently, the contemporary artist Banksy has used stenciling in the service of powerful social commentary. His informal art on city walls tackles subjects such as war, consumerism, and imperialism. You likely remember, in fact, his *There Is Always Hope* image of a child releasing a heart-shaped balloon. Some believe it symbolizes her loss of innocence, yet it still encourages viewers to not lose hope.

And, of course, the Beatles forever transformed much of the world. Music, human rights movements, attitudes toward mind-altering drugs, a fresh image for young men, and fashion were never the same after John, Paul, George, and Ringo. It's no wonder that autocrats eagerly try to destroy such artists. They shake things up.

Thus, by embracing the arts, we celebrate the diversity of models. But we also help ourselves consider, adapt, and innovate. Hopefully this results in societies that are kinder, more reflective, and ready to evolve.

Beyond Models

So far, we have seen some of the practical benefits of flexibly adopting different models. However, some people experience far deeper and more personal benefits from moving beyond models altogether.

We see this in an ancient Zen story, or "koan," that encourages people to give up their commonsense fixations on conventional models. In this koan, the student Fayan goes on a pilgrimage. His teacher Dizang soon learns that Fayan knows neither his destination nor the purpose of his journey:

> Dizang asked Fayan, "Where are you going?"
> Fayan said, "Around on pilgrimage."
> Dizang said, "What is the purpose of pilgrimage?"
> Fayan said, "I don't know."
> Dizang said, "Not knowing is most intimate."

The key insight, encapsulated in the phrase "Not knowing is most intimate," reflects the idea of a departure from fixed concepts and instead embracing uncertainty. It moves away from models and encourages an attitude of open-mindedness. This does not imply that we should throw out commonsense thought. Rather, it suggests that letting go of our preconceptions, or "not knowing," moves us closer toward experiencing reality directly, thus establishing a deeper, more "intimate" connection with the world.

CHAPTER FOUR

CONTEMPORARY COMMONSENSE MODELS OF HOW THE MIND WORKS

We've examined several commonsense models depicting the workings of the mind. None of them rest on a solid scientific footing. They often stem from sources such as community leaders, our biology, media, therapists, cultural norms, peers, and more. Despite their lack of a strong empirical basis, they seem true to certain people.

These models tend to quickly become ingrained in our lives. Like Hippocrates's humors, they sometimes go unchallenged for generations. Thus, they become part of the fabric of our culture.

For instance, as we saw in the preceding chapter, our enduring commonsense views of dissociative identity disorder are heavily shaped by social learning. Thus, one could argue that multiple personalities are the result of our social connections.

Researchers such as Elsa Ermer[1] believe that making commonplace inferences about thinking is a necessary part of these social relationships. In fact, some researchers (perhaps most notably Alan Leslie,[2] as well as David Premack and Guy Woodruff[3]) have described how, starting from a young age, we all develop these commonsense models, or what are sometimes called *folk theories of mind*. Using such commonsense ideas, we often attribute others' (or our own) actions to things and processes that we might only think exist in their (or our) brains.* These include inborn stubborn streaks, unconscious conflicts, hidden motivations, and unique spiritual, astral, or hypnotic states. In some circles, commonsense models even say that spirits, rather than our fingers, move the planchet across a Ouija board.

Modern commonsense models can also be used to describe emotions. For example, some people think of fear as being like a fluid that flows through the human body—one with certain physical properties, such as

---

* In other words, we project causes onto phenomena.

temperature and pressure. We communicate this when we say "they had a burning fear," or "their fear made them choke up," or someone "bottled up" their fears, or that their fears "came out under pressure." This is part of a larger *fluid theory of emotion*, where we might think of emotions as though they were hot liquids that sometimes boil over or stream out of us.

Even though emotions are not themselves liquids, thinking in terms of such pretend fluids can help us anticipate and influence people's behavior. For example, we might use the idea of "fear bubbling up" to predict someone's actions or to try to comfort and reassure them. But we can also think of this as a mental shortcut, the same way that we picture atoms as miniature solar systems, even though we now know that they are not.

We have seen that another modern commonsense model describes human emotions as though they were objects. We use this model to explain that our fears "resurface" (perhaps like a whale in the ocean) after self-exploration. We sometimes say that our emotions "bend us out of shape" and perhaps "push us" to act unwisely. In so doing, we think of fear as an object with material characteristics, like ropes and pulleys, crankshafts, and other things that we see around us.

Our Bodies as Architects of Our Commonsense Models

The habit of thinking about emotions as if they were fluids or recognizable tangible objects shows how much our physical experiences shape our mental models. Phrases such as "a weight on my shoulders" are far more than colorful language. They are important ways of describing our internal states such as moods, each of which is firmly anchored in our own physical sensations.

Anxiety, for instance, often presents as chest tightness. Perhaps this encourages us to think about fear as a constricting force. Likewise, think about the physical warmth from an embrace. It's suddenly not too difficult to understand why we might think of, or model, love as "heartwarming." We use the literal warmth of the moment to describe our affections.

Such connections between our physical sensations and the ways we describe the world suggest an interesting notion: Why not consider our bodies to be extensions of our minds? They greatly affect the way we model, comprehend, and describe the world. In this sense, our bodies are not just containers for thinking but active participants in our thought process. Looking at things like this, our bodies are an important part of mind itself.[4]

Invisible Puppeteers

There are many contemporary commonsense theories that involve unconscious "hidden forces in our head" thinking. For example, some people think that friends can't always tell us why they do things because their concealed turmoil influences their decisions. They might wonder if a hidden unconscious conflict causes them to prefer one friend over another.

Or think about a soldier with a mental illness called a *conversion disorder*. Just before battle, this physically healthy man can't move his arm. It is not clear why he can't move, but because he can't, he is removed from the field, sparing him from harm.

Because the soldier doesn't think, "I don't want to move," and he denies wanting to keep his arm still, we naturally look for something else to account for this seemingly puzzling "illness." Some people explain this by imagining unconscious and protective forces inside the soldier. These emotional entities supposedly paralyze the soldier to help him avoid war—but without him being aware of them.

While such models can be useful, the philosophical stance of "eliminativism" argues that these concepts do not necessarily reflect what really happens in our brains.[5] In fact, eliminativism suggests that these so-called hidden entities might not be the "currency" of an unconscious mind. Instead, they are perhaps just ideas that we use to make sense of human psychology.

Commonsense ideas aren't limited to unseen forces, of course. Many of us believe that insects, like humans, consider information and *decide* to move in one direction or another. But insects don't have the neurological sophistication that would enable them to make decisions in the same way humans do.

For instance, most people don't know that bees can recognize human faces in photographs.[6] This is quite a feat for an insect with a brain that's half the size of a gnat. However, this doesn't mean that bees "know" us and think, "Hey, there's that guy again." They likely are only reacting to patterns. Consequently, while it is sometimes tempting to take a mental shortcut and attribute human knowledge to insects, that would make little sense.

As we've seen, another commonsense model tells us that the spirits of dead people move a Ouija board planchet. Others, pervasive in many cultures, tell us that carrying fetishes—small statues of spirits—will protect us. Think of all the religious statues people put on their dashboards, as well as the beautiful little protective figures many African cultures prize. In some cases, these probably do protect people. Although I can't refute the religious conviction that this is because of divine intervention, my own belief is that they work by temporarily reminding the person carrying the fetish to be more vigilant and careful.

We have seen that each of these models can be useful in one way or another, even though they are often not based on science. For this reason, it might make little difference, from a purely practical standpoint, whether such models are accurate.

Imagining a Divided Mind

One widely held commonsense model of the human mind is that it has two parts that "think"—one conscious and the other unconscious—each with its own unique qualities.

Certainly, there are many things that we do, and that our brain processes, of which we are unaware. We all occasionally go on "autopilot," where we arrive at a destination but have little idea how we got there. We have made all the starts, stops, and turns without really thinking about it. Further, often when we blink or scratch an itch, we are not aware that we're doing it. But the idea that our minds have two separate parts that somewhat independently reason, remember, and have desires—one part of which we are fully aware of, the other not readily accessible to us—can be difficult to defend.

This supposed separation of the "conscious" and the "unconscious" thinking portions of our mind can also encourage some to imagine differences between them. For example, they might believe that our conscious mind is logical and our unconscious thinking mind is not, or that our unconscious mind has access to wisdom and insights that our conscious mind lacks.

As we explore these two fanciful minds, we could eventually imagine that we have discovered something important about ourselves. In fact, however, we could only be concocting more details about a model that might not be either true or useful.

It is especially tempting to use commonsense models of an "unconscious mind with thoughts" to explain people's irrational, selfish, or even criminal actions. For example, a psychologist in the early 20th century could have believed that a man did something awful to satisfy an unacknowledged desire to be violent. But how is this different from claiming that a rape was the result of an impulse planted in the man by the Devil—or that it resulted from the unhealed, suppressed memories that he suffered as a child?

What would happen if we didn't divide our mind into a conscious and an unconscious? For starters, our view of ourselves would have to change in some uncomfortable ways. We would have to accept that our irrational, self-serving, or criminal behavior reflects not some hidden alternate or suppressed personality but our *normal* self. We would need to take responsibility for the selfish or harmful actions that we attributed to a raging id, the Devil, or some troublemaker other than ourselves.

Our mistaken process of inferring our own inner workings can mirror how a space alien who had learned to read a clock might try to explain a wristwatch. If we look at the outside of an analog watch, we see numbers and hands that give us information: the time. The space alien observing that watch would perhaps wonder how the watch thinks about time, or how it knows to advance its second hand exactly once per second, or why it wants to tell time in the first place.* But the watch doesn't think or know or desire anything. And to display time, it doesn't need to do any of these things.

The Unconscious Mind as Sneaky Storyteller

There are many popular theories today that discuss the idea of "self-deception." Each invites us to think about how we deceive both ourselves and others.[7] Remember Antonio, for example, who seemed to use hypnosis to give some shape to an otherwise unseen portion of Anna's mind. We

---

* Perhaps you've heard the joke in which two philosophers argue about humankind's most amazing discovery. The first philosopher insists that it's $E = mc^2$, which she describes as "the universe's most elegant and important equation." The second nods and says, "I agree that it's an important equation. But you know what's much more amazing? The thermos." "The thermos?!" the first philosopher half shouts. "Why?" "Well," the second philosopher says, "it keeps hot drinks hot and cold drinks cold." "So?" the first philosopher demands. "What's so amazing about that?" The second philosopher leans close and whispers, with awe in his voice, *"How does it know?"*

saw that "Sarah" emerged and had a handy explanation for Anna's struggles with weight. However, "Sarah" was not, in fact, the savvy separate identity she portrayed herself to be. She didn't even have the right explanations for Anna's weight problem.

Just like we saw with Anna, some self-deception theories push the idea that an unconscious mind acts like a playwright.* Our fictional friend Sarah was supposedly the scriptwriter and gave Anna a compelling, yet still misleading, story line. In a dramatic interpretation, Sarah could be mistakenly likened to Shakespeare, secretly writing stories of past harassment for Anna's "conscious" self to consume.

With Sarah and others, the unconscious mind's alleged goal is often to deceive the conscious self and mask the true reasons behind one's actions. For example, some self-deceptive "Sarahs" might slyly suggest that Anna's generous acts come from the heart and not from a greedy wish to climb social ladders. So yes, we can easily conceive of a hidden entity, like Sarah, who convincingly fools Anna (and others) into believing in her generosity.

Thus, one particularly intriguing area that some self-deception theories promote is when they suggest that there is a thinking "unconscious" mind that deliberately spreads misleading information about the *true* reasons we do things. This is a bold move, given our often-poor grasp of what causes human behavior. It begs the question: How could it be that a small, hidden nook in our brain harbors exclusive insights into our actions?

Adherents to some of these theories also sometimes assert that we must resist our unconscious and destructive impulses, such as the orders that a "Sarah" would issue. It feels like there is an inner tug-of-war between our darker, hidden side and the values we've learned.

This model of both conscious thoughts and an all-knowing unconscious mind sometimes splits our mind into two factions along moral lines: a conscious part that likes to be good and an unconscious portion that's itching to be the rule breaker. ("Sarah," of course, was a helpful exception to this troublemaker label.) To address this, some approaches to therapy, rooted in this framework, aim to inject morality into our unconscious selves. They sometimes encourage telling patients to look deep inside themselves to find

---

* We will learn about a similar idea shortly, called the "left-brain interpreter," where a portion of the brain seems to strive to form coherent stories even when it has little or no information on which to base them.

their "true" motivations. The dream is that, by unearthing the secret and selfish reasons behind their actions, they will become better individuals.

In my view, and some will reasonably disagree, there is no convincing proof that our minds are divided on a moral battlefield. Even so, sometimes these methods work wonders.

The Risks of Over Relying on an Imagined Inner Conductor

The notion that we have an unconscious mind that conceals the true reasons for what we do is seductive, isn't it? But it is also fraught with danger. It can be an appealing model because it encourages deep, sometimes even calming attempts to introspect. It whispers, "Don't worry if you haven't figured yourself out yet; you have a hidden guide within you who will soon show you the way." Some people then get into self-exploration, meditation, psychotherapy, or even philosophy. After this, they might think they've tapped into a hidden inner source of wisdom. Anna seemed to do this. However, after having discovered that Sarah was largely a product of her imagination, we can hope that later the resourceful Anna thought more critically about supposedly having illuminating beliefs stemming from an unconscious mind.

But here's another problem: If we think that we have an unconscious mind that holds the keys to self-knowledge, we might become novelists ourselves, fabricating even more elaborate fictions. Anna went all out on this with her role in creating Sarah. *That is, in an overzealous attempt to uncover "deep truths in our minds," we instead construct or reinforce more delusional beliefs, or even false memories.*

Therapists like Antonio need to tread carefully here, especially when trying to find "repressed" memories or truths hidden away.

For instance, Anna and Antonio teamed up to produce Sarah, who had us all believing that her overeating was because of childhood ridicule. It sounded like a revelation. But as we learned, her memories of being mocked and being embarrassed were only a story coming from imagination and suggestion. They turned out to be make-believe rather than true insight.

Abandoning Unconscious Scheming

Alfred Mele[8] has his own down-to-earth idea of self-deception that avoids many psychological smoke-and-mirrors. He doesn't use the idea of an unconscious as a schemer behind the curtain. Instead, he offers a simple, more straightforward model. To do this, he turns to the simple idea of cognitive bias.

We can think of cognitive biases as little mental shortcuts. Better yet, think of them as some of the brain's "life hacks." They help us process information more quickly. Some can be useful, but they can also mislead us. The example of cognitive "confirmation bias" is a nice example. Recall that it is a sad tendency to focus on evidence that confirms what you already think you know. Perhaps you think that chocolate is a health food. With confirmation bias, you would likely pay more attention to stories that praise chocolate's health benefits than to those that condemn its sugar content. The same goes for opinions on politics, or even arguing about whether a movie is Oscar-worthy or a flop.

Simply put, we are attracted to information that pats us on the back and supports what we think we know. We run away from knowledge that threatens to shake our mental comfort. Think about what can happen if we are convinced that we will get a promotion at work. In his model, we misinterpret even tiny clues about our coworkers' abilities. We would be like pretend detectives wearing blinders. With this, we would only pay attention to information that supports our self-image as a rising star and ignore data that show how good other people are.

Compared to some other ideas, Mele's model is also often far more helpful in therapy. First, it's a lot easier for a therapist to put into practice. If patients are shown that they already have an unfortunate habit of cherry-picking the information that supports their beliefs, then they can work on changing it. With this, they might eventually become better at considering different sources of information. Obviously, this can be a more useful model than imagining the far more abstract, and supposedly deceptive, antics of an unseen and persuasive force.

The Mysteries of Automatisms and Dissociation

So far, we have seen examples of multiple hidden, unconscious scenarios, each supposedly explaining why people act in certain ways. These models come to the forefront when people try to explain human activities called *automatisms*: things we do without any sense of intention.

Automatisms[9] often show up at religious events, when a spiritual being seems to communicate through a person by "causing" them to speak in tongues—usually unintelligible noises called *glossolalia* that they believe are inspired by God. This can include such diverse groups as people who practice voodoo, shamanism, paganism, and Pentecostalism. Some see this as evidence of hidden spirits inside people's heads; others take these automatisms as signs that the speaker has allowed the Holy Spirit to enter their body.

Automatisms also show up when people play with a Ouija board, seeming to allow spirits to move the planchet under their fingers and spell out information from another world.

Some explain automatisms using the model of *dissociation*, in which people experience a profound separation between their conscious thoughts and their bodies—or between their conscious thoughts and their actions or speech.

Dissociation can occur during hypnosis, when people believe themselves to be "under the power" of the hypnotist. For example, when Anna first closed her eyes in response to her therapist's suggestion, she experienced her eyes closing—but she did *not* sense herself willing her eyes to close.

Others see this loss of feeling in control over one's body as the blurring of boundaries—either of the self or the mind—between the therapist and patient. When this happens, the therapist's instructions can be interpreted by the patient as having come from inside themselves.

Clinical Views on Dissociation and Mental Health

Many mental health professionals consider dissociation to be a hallmark of a variety of mental disorders, including dissociative identity disorder, some memory concerns, and occasional feelings of being detached from (or "outside of") one's body.

Some contemporary therapists, following more recent psychoanalytic versions of Freudian theory, see dissociation as involving long-standing painful memories or relationships that have not yet been integrated into people's current ideas of themselves. The resulting "split-off" discomfort is therefore something that seems foreign to the thinker.[10] Some clinicians help their clients address this by encouraging them to see the split-off pain as an unhelpful model—and instead as a response to traumatic events that occurred earlier in their lives. These clinicians believe that this perspective can result in insight and improvement by helping a client to integrate their previous experiences into a more inclusive sense of themselves.

Lay Explanations of the Unconscious

When we cannot make sense of someone's (or our own) actions in terms of conscious, deliberate intent, laypeople tend to default to one or more commonsense, poorly supported models. So do some scientists, researchers, and mental health professionals.

Put more simply, when we don't know exactly why we (or someone else) did something, we often default to an explanation that looks, sounds, or feels right based on the models that are widely available to us.

For instance, suppose that you leave your house to go to dinner with your sister but end up at a clothing store instead. You might attribute your changed destination to unconscious motives (*I'm unconsciously angry at my sister for the way she acted last week*, etc.). Meanwhile, you ignore or dismiss the effects of how much you like the attractive sweater you saw in the store window and your dislike of the loud and expensive restaurant she selected.

As for the other supposed entities in your head—your "unconscious" beliefs, a trance, a demon, etc.—think back to your childhood. Did you ever have an imaginary friend? When you did something that upset your parents, such as break a glass, did you ever blame an imaginary person for the transgression? Many of us became comfortable with such imaginary entities long before we became comfortable with holding a pen. But we also see that as we age, our attraction to make-believe, default, and comfortable explanations eventually takes more complex forms. Our "imaginary friends" become inner critics, intuition, or muses.*

---

* Also note that dissociation is, in a way, part of our outside world. It reflects how we are taught to define the limits of ourselves, and distinguish ourselves from other things.

We have learned that while commonsense models are convenient, they often persist because they resonate with personal experiences and cultural beliefs. It is crucial to recognize both their limitations and the potential for distortion they introduce into our understanding of ourselves.

CHAPTER FIVE

SCIENCE AND MODELS OF THE MIND

C an we do better than our commonsense models of the human mind? Also, can we do better than models used by professionals that have practical value but sometimes limited scientific backing?

Yes and yes.

Let's begin in the year 1620, when Sir Francis Bacon published his *Novum Organum* ("New Instrument").[1] In this book, Bacon described a new method of learning that laid the foundation for the scientific method. He advocated a systematic observation of the world, from which we can develop general laws.

This work changed the world and left much guesswork behind. He offered a better way to sort through a complex world and shifted attention to careful observation and study.

Because of Bacon, it soon became commonplace for scientists to tackle questions by following this sequence of steps:

1. Identify and describe the problem.
2. Search for facts to help solve the problem.
3. Make reasoned guesses—i.e., hypotheses—about a possible solution.
4. Design a study to test the hypothesis.
5. Carry out the test and harvest the results.
6. Critically examine those results.

These steps gave a structured plan for solving problems. Now people had a reasoned map that guided them through new territory.

By showing people how to reach conclusions based on real-world experimentation and solid analysis, Bacon helped start the Scientific Revolution. With this, he triggered unparalleled advances in science, medicine, and quality of life. In a short time, people could turn some formerly fatal illnesses into manageable, or even curable, conditions. It was the start of a method of thought that would eventually land humanity on the moon.

Once Bacon's work became widely accepted, it was easy for many people—especially scientists and researchers—to flatter themselves by assuming that his method of reasoning reflected how the human brain naturally worked. Put another way, once people got used to running Bacon's software in their heads, they began to imagine that the scientific method was ingrained in their mental hardware.

But of course our brains often don't work that way. Not everything about the scientific method comes naturally to human beings. We each must learn it. And this takes time and effort, just like learning to play the trumpet.

In fact, much of our thinking is totally unscientific. How many of us knock on wood, cross our fingers, don't walk under ladders, wear lucky underwear, or shudder when a black cat crosses the road in front of us? These habits have nothing to do with logic and science. Yet they are quirks that remind us that science isn't necessarily a model that we naturally use to perceive the world.

Measurement Pitfalls

While Bacon's scientific approach was a huge advance, it had its own limitations. These parallel, and highlight, the limitations of all scientific models. As we will see, models aren't perfect, and none of them are more than approximations of reality.

One of these issues is the problem of measurement—of things, events, and actions. To make any measurement, we need something to measure, something to measure it with, and someone or something to do the measuring. This is obvious.

Let's start by looking at the measurements people use to understand the world. Virtually without exception, each one is based on their earlier ideas of what we thought, or were told, was true. But here's the catch: These tools are not necessarily the best way to capture the full picture. Sometimes they can even strengthen our earlier misconceptions or biases because they were made to look only for certain things. This can be like looking for lost silverware using a metal detector that totally misses stainless steel.

One example is early IQ tests. They were often based on now dusty beliefs about intelligence. Some overlooked parts of intellect, which included both creativity and social smarts. They might even unfairly lean toward certain cultures. It's easy, of course, to understand how imperfect measures such as these were often like self-fulfilling prophecies. They commonly reinforced a narrow view of intelligence and discouraged broader and more complete understandings.

Just as these tests have had some major upgrades, Anna and her therapist learned that their first assumptions about human psychology were outdated solutions for new problems. They were missing a lot. For instance, we saw that Antonio gave Anna a hypnotizability scale. This sounded very scientific. However, because it only looked at what *she* did, it played into the idea that the only way to study hypnosis was to look in her head. As we will later see, this narrow perspective failed to capture the rich complexity of hypnosis because it did not focus on what happened in Anna's surroundings.

We can see that in other areas of science our measurement tools can profoundly impact our findings. This is exemplified in physics by the classic double-slit experiment, which has shown that the very act of measuring or watching light can change whether it seems to manifest as waves or particles. In these double-slit studies, using a detector to simply observe the path of light seems to make that light appear as particles rather than as waves. But when we use our eyes to look at the pattern they make after passing through slits, the light can appear as waves.

Observing human beings creates a similar set of problems. Psychological observations used in research can also impact the topic of a study. For example, think about when subjects know they are being watched. What if you were on a diet and you know a researcher is watching every bite you take? You can't tell me you wouldn't eat differently.

Biases and Beliefs in the Scientific Method

The scientific method is also subject to experimenters' biases. There is nothing in the method itself to prevent people from conducting studies that tend to reinforce their existing mistaken notions. People who believe the world is flat or think that aliens created landing strips in Nazca, Peru, still somehow find ways to "prove" their beliefs. Like this, researchers who steadfastly believe in ancient astronauts have designed studies in ways that appear to show that such imaginary entities really lived long ago.

The scientific method, by its nature, also preserves a view of the world as separate, discrete objects. It divides and analyzes, often with profoundly valuable results. However, when using it, researchers often ignore other views of the world that see systems as larger, integrated wholes. What if you have a hammer and you ignore your pliers, wrenches, drills, and screwdrivers? You will only focus on pounding nails and will be lost when you are faced with a problem that needs a twist.

We should not misunderstand the scientific method as a path toward absolute and "final" truths. Instead, I think of it as something that refines our thinking. Science only gives us models, each one subject to replacement as more is learned.

Science Adapting to a Changing World

In essence, it is good to think of our models, both those based on science and those based on common sense, as more like works in progress, subject to change, rather than being static and firmly planted. By piecing all this together, it should be clear that all social and technological changes, and many other advancements we haven't discussed, reshape our thoughts about the world and consequently impact scientific objectivity. We see, therefore, that some of science's limitations aren't just science problems. They are instead entwined with the dynamic nature of human experience.

Behaviorism versus Commonsense Models

Behaviorism is a school of psychology that relies heavily on the ideas of natural selection and observed action, rather than on unconscious forces or inferred motivations. It is a strong example of a scientific approach to learning about ourselves.

If you were a "just-the-facts" detective, you might decide to follow the lead of behaviorists. You would only respect the hard evidence, that is, things that you see people do. You wouldn't spend time guessing about how someone feels, or what they think, unless you could observe those things yourself. Stepping away from speculation like this can be a welcome breath of fresh air.

Suppose you have friends who never offer to pay the dinner bill. You might wonder if they don't value or like you, if they are concerned about financial problems, or if they dislike eating out. If you were a behaviorist, you would perhaps not even consider asking about these things. You would instead focus on patterns in their behavior for clues to why they behave as they do.

We see here that because behaviorism stresses studying observable actions, it avoids much of the guesswork about unseen psychological events that had long plagued psychology. It therefore does not make some of the mistakes about psychology that would otherwise prevent us from learning the truth.

Ivan Pavlov, who heavily influenced behaviorism, is acclaimed for his theory of classical conditioning. In experiments with dogs in the 1890s, he rang a bell shortly before the dogs' feeding time each day. He discovered that, after enough repetitions, the dogs would reliably salivate on hearing the bell, before they were exposed to any sight or smell of food. The bell alone (the "conditioned stimulus") caused salivation (the "conditioned response"). This makes sense under the umbrella of natural selection, because dogs salivate to prepare to digest food shortly before they eat it.

You can think of it as being like when you hear your phone's ringtone, and you feel excitement before even looking at the screen. That signal means a reward might be coming. Perhaps it is a message from a friend, a job offer, or a business success. Although simple, it can be a powerful way to understand some behavior.

Pavlov could almost always predict conditioned salivation by controlling the dogs' surroundings. He didn't have to find out what the dogs were thinking or feeling; all he did was study their observable behavior. This was a groundbreaking moment in psychology.

In the early 20th century, behaviorist psychologists such as John B. Watson continued Pavlov's revolution.[2] Like Pavlov, Watson largely turned away from the study of thoughts. Instead, he examined observable behavior in both humans and animals. This included speech, voluntary muscle movement, and involuntary physical responses.

Over time, as technology improved, behaviorists were able to measure many kinds of responses, including electrical impulses in the brain. And so, we eventually began, in limited ways, to observe and measure human thinking—with the caveat that our observations and measurements might affect that thinking.

Eventually behaviorism revolutionized psychological treatment. It led to many of the first successful cures for eating disorders, depression, self-mutilation, anxiety, learning difficulties, and other problems.

Let's think about how a behavioral treatment could have helped Anna. Instead of delving into a supposedly unconscious thinking mind, a behaviorist could give her tools to change her eating habits. It would be straightforward. She might be coached to avoid unhealthy snacks, track her diet, and keep a record of her weight changes. Little of this would be based on guesswork about what was going on inside her head.

It's important to pause here and notice the key takeaway: *Some of psychology's greatest leaps forward occurred when professionals stopped trying to chase down supposed hidden motivations, entities, and memories.*

Behaviorism, however, doesn't dismiss the study of thinking. Behavioral psychologist B. F. Skinner saw thinking (which he described as "covert behavior") as a method to mentally run simulations and test out various courses of action to see where they will lead. We can do this, for example, by thinking about which argument might most persuasively encourage a store owner to give us a discount. I weigh different possible courses of action much of the time, and you do too.

Nevertheless, Skinner correctly felt that useful research on these simulations would only be possible when they could be physically observed in some way—for example, with brain scans. Put another way, thoughts could be studied only after we had the technology to better view the brain in action.

The concepts of natural selection influenced Skinner. In his explanation of how thinking has helped us succeed as a species, he wrote, "Covert behavior [such as thinking] has the advantage that we can act without committing ourselves; we can revoke the behavior and try again if the private consequences are not reinforcing [i.e., rewarding]." In Skinner's view, our thinking made humans more competitive by allowing us to mentally test different potential actions and "have in mind" their likely results before carrying them out.

Skinner therefore studied humans as competitive biological organisms who must constantly cope with the demands of evolution.

Let's stop and note another key distinction here: *Behaviorists' approaches to studying people reflect the knowledge that people's conclusions about why they do what they do are often wrong.*

As we are beginning to see, rejecting our "commonsense" ideas, conclusions, and reasoning can sometimes be essential for uncovering truth—or at least for finding more useful ways of examining ourselves and the world.

This is uncomfortable for most of us. We love our own explanations. But we have seen that our common sense is often flawed. And so, being a behaviorist can help us see the world as it is, rather than trying to peer through opaque curtains or believing false stories that we like to imagine.

Leaving a Flat Earth for Quantum Mechanics

Such principles aren't just true for studying the human brain. Physicists and astronomers have a long history of disproving and discarding commonsense scientific theories. Some of these theories were incorrect but nevertheless useful; some were neither.

You're already familiar with some of these theories, which include:

- The Earth is flat.
- The moon is a light source.
- The sun revolves around the Earth.
- Heavy things fall faster than lighter things.
- Objects either move or are at rest in an absolute sense, rather than in relationship to their surroundings.

Here is another, less widely known example: the ether theory.

This was the common belief in the field of physics that "luminiferous ether" was an invisible and somewhat mystical substance that helped light waves travel through space. It was a convenient explanation because people couldn't conceive of how light could move through a vacuum. But in the late 19th century, researchers found that there was no evidence of luminiferous ether. This finding shook things up and pushed science toward more accurate models[3] that stood up to scrutiny.

Further, as recently as about 100 years ago, people thought that light was strictly a wave. Light does have properties like those of waves; for example, it bends around obstacles, like waves in a pond curving around reeds. And when lights from multiple sources intersect, the waves both cancel and reinforce each other, just like waves on a pond.

Einstein speculated that, while light did show properties of waves, it *also* had characteristics of particles. He theorized that these particles—later dubbed *photons*—knocked electrons loose from metals, possibly like billiard balls striking against one another,* in what came to be known as the *photo-electric effect.*

Einstein turned out to be right. Over time, many experiments disproved the commonsense view of light as *just* a wave. Instead, as we know, depending on how we observe it, light appears to be something very strange and nearly impossible to conceptualize: both a wave *and* a particle.

Just as some of Anna's mistaken ideas about her childhood were soon replaced, we see here that the wave theory of light was superseded by more comprehensive thinking. Both advances gave far more detailed and complete knowledge of reality.

Here's another commonsense view about light that physicists have demonstrated to be false: Most of us think that light travels much like an arrow or a bullet, but much faster—at 186,000 miles per second. The speed is correct, but it turns out that the arrow and bullet metaphors are not.

As many experiments have demonstrated, light moves through space in relation to *anything* material at 186,000 miles per second.† Light from the sun strikes the Earth at 186,000 miles a second. But what if you are in a spaceship traveling toward the sun at 100,000 miles per second? Common sense says that the light should be coming at you at the combined speed of the light and your spaceship, or 286,000 miles per second. But it doesn't. It comes toward you at 186,000 miles per second. Using our commonsense understanding of light, we might wonder how that could possibly happen.

The opposite is also true. If you and your spaceship are speeding away from the sun at 185,000 miles per second, then our commonsense view tells us that the sun's light would strike us at 1,000 miles per second. But, again, it doesn't. It strikes us at 186,000 miles per second. In this case we would likely think that 186,000 - 185,000 = 186,000.

When it comes to light, our commonsense ideas turn out to be just as imaginary as bodily humors and demons.

Equally imaginary is our concept of space and time as separate things. Einstein theorized—and experiments later verified—that space and time are best viewed as a single phenomenon.

---

* Photons, of course, don't really behave like billiard balls, as they have unique quantum properties not seen on a macroscopic level.

† Light does slow down fractionally when it moves through objects, such as water or glass. In highly unusual conditions—i.e., in a laboratory, in sodium atoms cooled down to almost 460 degrees below 0 °F—the speed of light has been reduced to 38 miles per hour.

Later work in physics revealed that what we think of in a conventional sense as space and time exist only as human creations. They are necessary models that we can easily comprehend. However, they are not accurate portrayals of reality.

The development of quantum mechanics, which described previously unseen subatomic particles, broadened physicists' views of the world. These were different from previous theories of classical physics in that they described events that didn't otherwise make rational sense, such as the unchanging speed of light in a vacuum. These laws do not necessarily follow our commonsense ideas. They are useful to our conceptions of the world, but extremely difficult to comprehend.

For example, try thinking about a subatomic world in which things suddenly disappear and then, just as suddenly, emerge someplace else. This is difficult to visualize or accept. Yet it has been demonstrated, over and over, to be true.

This broadened perspective—using ideas not chained to our old commonsense models of the world—was a major scientific advance. It helped us, for instance, develop revolutionary new materials and is now enabling us to design computers many times faster than those we have today.

The Einstein/Infeld Watch

In 1938, Albert Einstein and Leopold Infeld[4] described our limitations for comprehending reality using this metaphor:

> In our endeavor to understand reality we are somewhat like a man trying to understand the mechanism of a closed watch. He sees the face and moving hands, even hears its ticking, but he has no way of opening the case. If he is ingenious, he may form some picture of a mechanism which could be responsible for all the things he observes, but he may never be quite sure his picture is the only one which could explain his observations. He will never be able to compare his picture with the real mechanism and he cannot even imagine the possibility of the meaning of such a comparison.

This metaphor still holds true, even though we've come a long way in devising new scientific theories since 1938. Despite our advances, many of our limitations in knowing about the complex world remain.

In the 21st century, we can think of Einstein's example as comparable to trying to learn about the source code of software, but only by using the application. We can open menus, click buttons, and perhaps try to find hidden features. But without seeing the code itself, we would continue to have a superficial understanding of what's really going on. Unless we were programmers ourselves, we could only make rough guesses about how it, or reality, truly works.

As we have seen, even the newer models we use might not feel natural because our brains are better equipped for (and more comfortable with) working with concepts that align with common sense. With this, we again seek simplicity. As a result, we often choose poorly supported ideas instead of the truth.

Anna's case shows this concept perfectly. Her so-called "memories" of being teased during childhood seemed believable. They fit into common-sense notions, particularly about psychological reasons for putting on a few extra pounds. However, her memories were not true but only a convincing mental model that she, and her therapist, found easy to believe.

Perhaps our species has survived and thrived partly *because* its thinking has not been tethered to truth but to whatever has promoted the ability of individuals to survive, reproduce, and pass on their genes. Accurately pondering the true nature of reality was far less important than coming home with a mate. Knowing this, it is obvious that how we see the world often reflects what best keeps our species going rather than what is true.*

We thus see that, when it comes to the human brain, there is a lot to consider. But for starters, let's embrace a more scientifically grounded view of how our brains work. However, while doing this we still need to see beyond our scientific habit of dividing reality into tiny pieces, a habit that obscures the larger picture.

Hopefully this will broaden our knowledge about ourselves within the world. As we will see, this approach can often free us from the self-imposed limits of old models, perhaps much like Anna did to lose weight using hypnotherapy.

---

* This would largely explain why so many people quickly (and seemingly intuitively) take up beliefs that are clearly and demonstrably false but are promoted by a great many very powerful people.

Part II

LETTING GO OF OUR FICTIONAL MINDS

We would rather be ruined than changed. We would rather die in our dread than climb the cross of the moment and let our illusions die.

—W. H. Auden

CHAPTER SIX

THE MYTH THAT OUR MINDS RARELY MAKE THINGS UP

Most commonsense models of how the mind works imply that we usually see the world correctly and then use that information to make the most logical, useful decisions. Anna probably felt this way—even *after* she realized that her childhood enemies were imaginary.

This line of thinking allows us to envision an idealized self, one that can consistently guide our lives based on a rational analysis of information.

In essence, we think that logic is our default mode. We don't believe that we make things up.

In fact, however, we have seen that we make up stuff much of the time—and what we call logic is often anything but logical. Let's look at a few examples.

Picture yourself sitting face-to-face with someone who has a bowl of M&Ms hidden under the table. They have told you that 80% of the M&Ms are yellow and 20% are green. They will choose one M&M at random, hold it in their closed fist above the table, and then open the fist. Your role is to guess the color of the M&M before you see it. If you guess right, you can eat the piece of candy. And you like chocolate.

To get the most M&Ms, you have several possible strategies:

- You can say "yellow" or "green" randomly and hope for the best.
- You can guess based on your intuition.
- You can always choose yellow, knowing that you will be wrong about 20% of the time.
- Since 80% of the M&Ms are yellow and 20% are green, you can say "yellow" 80% of the time and "green" 20% of the time.

Which strategy will earn you the most treats?

Fish figure out this kind of thing quickly (though with fish food for their reward rather than chocolate). In experiments, fish soon pick the most frequently occurring color almost every time—and for that reason, in this example, they would get rewarded about 80% of the time.[1] This is both straightforward and logical.

We humans are typically much less effective.[2] In experiments, most people say "yellow" about 80% of the time and "green" about 20%. Perhaps we do this to replicate the 80/20 ratio, which some think is a logical move. But it's not the most successful strategy at all, and it often results in a success rate of less than 70%—notably worse than what fish routinely achieve. *We might think that we have relied on a logical, time-tested strategy. However, here it is not. The fish win!* (To be fair, mathematicians, actuaries, bookies, and other people who calculate odds for a living generally don't fall for this misunderstanding. They often have as much success as the fish.)

Our frequently poor decision-making skills, based on faulty logic, can be seen in our perception of the world, such as when we miscalculate odds. We just saw this happen with M&Ms. But also, many people are more afraid of flying in a commercial aircraft than they are of driving a car. In fact, though, statistically by far the most dangerous part of commercial air travel is the ride to the airport.

Order and Illusion

Michael Gazzaniga and other researchers[3] have shown that when others see (or are told about) a sequence of randomly selected events, they will often perceive in those events some pattern—even when no such pattern is actually present. For instance, with the M&M scenario, some people think that the M&Ms are selected in a way that is matched to their ratio in the bowl. But of course the M&Ms were chosen at random, and the only patterns were in the observers' minds.

Gazzaniga also discovered that the left side of the human brain routinely makes up "facts" and explanations. Sometimes these appear completely logical and reasonable—yet they are also verifiably false, just like Anna's childhood bullies.

Gazzaniga says that the left brain is responsible for an interpretive process that is the "glue that keeps our story unified and creates our sense of being a coherent, rational agent. To our bag of individual instincts it brings theories about our life."[4] Yet these theories are often untrue.

In fact, our view that we routinely make logical decisions is itself often an illusion produced by the brain's interpretive process. But why don't we see this? Before you answer, remember that nature would not necessarily have equipped us to identify our illogic, because we are built for fitness, not necessarily for discerning (or caring about) the truth.

In short, our almost unshakable belief that we are logical, like Anna who first accepted her false memories, is evidence of our brains' skills at storytelling.

Confabulation as the Mind's Mirage

We can see the left brain's faulty interpretive process at work with a phenomenon known as *confabulation*. This is the automatic filling of gaps in our memory with fabrications that we make up and accept as facts. Put simply, when we don't know something, our left brain might make up an answer.

Confabulation is common among people with brain disorders who have been treated by cutting the bundle of nerves that connects the left and right halves of their brains. Although this operation stops seizures in people with epilepsy, it also prevents information from moving between the left and right sides of the brain.

Many experiments have been conducted on people who have had this "split-brain" surgery. In some of these experiments, people are shown signs with various statements and commands. However, these signs are projected in such a way that they are only visible to the portions of people's eyes that are connected to their right brains. The visual signal goes from the eye to their right brain but never makes it to their left brain.

In most humans, the right brain can read, follow commands, and do a variety of other things, but it is not focused on making up explanations. As a result, when researchers[5] showed a subject a sign that said "Laugh!" the person laughed because their right brain read the sign and responded. Then the researcher asked, "Why did you laugh?"

Answering that question is a job for the left brain. But in these experiments, subjects' left brains had not seen the command to laugh, so they had no context for the laughter. The left brain therefore could only observe the laughing but could not understand the reason behind it. To subjects' left brains, it was as if they had watched a stranger laugh at something invisible.

But this did not stop the left brains from making up false—but reasonable-sounding—explanations. When the researcher asked, "Why did you laugh?" almost no one said, "Hmm. I don't know. Good question." Instead, people said things like, "You had a funny expression on your face," or "You reminded me of an old joke," or "A really odd-looking car just drove by outside."

In another such experiment,[6] a man who had received split-brain surgery was shown a sign that said *Walk!*, so he stood and walked across the room. But because his right brain could not "tell" his left brain about the sign, his left brain was unaware about why he had begun walking. When researchers asked him why he had gotten up and walked away, he explained, "I'm going to get a Coke." His left brain had reflexively (yet confidently) confabulated an explanation for his behavior—an explanation that was reasonable, believable, and incorrect.

Here is what is notable about these responses:

- They were all left-brain fabrications, made up on the spot.
- Almost every subject did this.
- People who fabricated something seemed to *believe* it and did not recognize its absurdity.
- Their explanations were made as if a person was drawing conclusions after observing the actions of another individual.

The Puzzle of Split-Brain Surgery

It's stunning that people who have undergone split-brain surgery often report that they feel no different than they did before their brain was partly cut in half. After all the changes that occur in their thinking, they perceive no differences in their reasoning skills, problem-solving abilities, emotions, or everyday thoughts.

What is even more intriguing to me about split-brain surgery is how skilled the left brain is in maintaining a continuing sense of having one "self." It pulls this off even after it has been partly isolated from its counterpart. It subjectively makes a feeling of a unified consciousness, perhaps not giving even one clue about its surgical separation from its other half.

This is interesting for the average healthy person who remembers that they made a choice based on a "gut feeling" or that they justified an earlier impulsive decision. This person's brain might have been working overtime to create a story that made them feel both unified and rational, even when neither of those things was true.

Confabulation beyond Surgery

Confabulation doesn't just happen in people who have received split-brain surgery. It is also common among people with strokes, Alzheimer's disease, brain injuries, brain tumors, and certain other brain illnesses.

In some cases, confabulation can be extreme. One woman[7] whose left side was paralyzed by an illness was completely unaware of the paralysis. She was also unable to recognize her left arm and leg as her own. When a researcher asked her to explain the presence of those limbs, she said calmly that there had been another person in bed with her—a little girl—whose arm had somehow slipped into her sleeve.[8]

There is a case of another man[9] who, after having suffered brain damage, began taking on the persona of whomever he was with. After he was with psychologists, he confabulated being a psychologist himself. After he was with a lawyer, he claimed to be a lawyer. When the researchers took him into a hospital kitchen, he later said he was responsible for preparing special meals for hospital patients. Finally, when they brought him to a bar, he thought he was a bartender.

Some researchers[10] have explored "false memory" confabulations about alleged childhood abuse. Typically this happens after counselors encourage patients to interpret normal events, such as not remembering much of one's early childhood, as evidence of abuse.[11*]

Of course, false memories are different than confabulation coming from a brain injury. First, Anna's false memories were perceived as more psychologically rooted rather than physical. This is because she had no brain trauma. Meanwhile, confabulation coming from neurological damage can have various physical causes. Knowing these differences helps us understand both types of errors: those arising from psychological issues and those stemming from neurological damage. Both are relevant to this issue.

Mental Tricks and Evolution

False memories can be important from an evolutionary point of view. They could be like mental safety nets, helping keep you calm and functional and making life seem predictable. What if, instead, your brain didn't fill in some

---

* Children are especially likely to confabulate false memories.

information? You would be second-guessing every small decision you made and be exhausted in short order. That's not a good recipe for survival.

Such likely necessary traits have probably been finely tuned over millions of years in our evolutionary history. A unified sense of self, along with a logical story, really helps when trying to survive and communicate with others. A believable and unified story can likely help, even if it is based on false information. This is not foolproof, of course. But, overall, this trait probably helps us get along in the world.

This hits home when we think about the story of Anna. Her brain produced a not-so-accurate story that perhaps helped her with weight loss. She didn't need to know the whole truth to benefit. And so, if someday you find yourself creatively filling in the blanks, it might be part of your brain's toolkit that helps you cope with the world so beautifully.

Confabulation beyond the Clinic

It's tempting to conclude that confabulation is a form of dysfunction that's only found in people with certain illnesses and uncommon psychological problems. In fact, as should be apparent to you by now, evidence strongly suggests that it's a reasonably common human trait. Confabulation is simply more obvious or absurd in people with brain impairments.

In one experiment[12] with normal people, researchers posing as sales associates in a store showed shoppers four nylon stockings. They described the stockings as four different brands from four separate manufacturers. They asked the shoppers to feel and stretch all four and then to pick their favorite. After each shopper made her choice, she was asked to explain her decision. Each one had no trouble choosing—and no problem clearly explaining her reasoning. However—as you have probably intuited—all four stockings were identical. *Every such explanation was a confabulation.*

Some years ago, one of my friends needed surgery for a detached retina. He was prepared for surgery by a blond, heavyset female nurse, then given an anesthetic that made him unconscious.

He regained consciousness midway through the surgery. After a minute or two he was fully awake and alert. However, he had only very limited sensory input. His healthy eye was covered, and through the eye being worked on, he saw only a blur of light. Still, he could clearly hear people speaking and moving around him. Perhaps based on these sounds, he later remembered that the operating room was small and that three people—the eye surgeon, the prep nurse, and a second nurse with similar build and features—were clustered around him.

After a few minutes, the surgeon sewed up his eye, said, "All right, it's done," and left the room. A few seconds later, the covering was removed from his healthy eye, and he looked around. Reality clashed with perception.

The operating room was spacious. The nurse who had removed the covering was a thin Black man. There was no second nurse.

After a moment of confusion, he caught himself and laughed inwardly. He realized that, in the absence of any visual information, he had confabulated (likely with the assistance of the anesthetic) the entire "experience" of those minutes of surgery.

Egos and Algorithms

Confabulation, and perhaps biased treatment of information, has caused a good deal of head-scratching, even for some seasoned mental health professionals. We likely have seen this in the decades-long turf war known as "clinical versus statistical prediction." This struggle has pitted some trained counselors such as Antonio against the logic of numbers. It is a battle that was ignited in the 1950s by psychologist Paul Meehl.[13] He stepped well out of many psychologists' comfort zones and showed that statistics and algorithms often (but not always) do better at evaluating patients than do well-trained mental health professionals. What a surprise for many! This included spotting mental health problems like anxiety, as well as attempts to deceive the mental health professionals interviewing them.

Think of the implications by imagining two arguing psychologists. One uses 40 years of training and experience to talk to the patient and learn whether that person is anxious. The other takes a different route. She doesn't even chat with the patient. Instead, she gives that person paper-and-pencil psychological tests, each based on mathematical algorithms.

Common sense often tells us that the experience gathered over decades would always destroy any algorithm. Well, many psychologists held, and still hold, to this idea. However, over half a century of study has shown that, with many psychological issues, the boring numbers work best. Sometimes by a lot. This is a knockout blow to many doctors' egos.

Gazzaniga's concept of a left-brain interpreter adds a new twist to this struggle. We know that it looks for patterns, makes inferences, and constructs models to help us interpret the world. Now, picture a seasoned psychologist with a storied career of good work. We might think that her brain would write a flattering conclusion: You're the expert. You see things that numbers can't grasp.

Well, sometimes that's true. And it's a nice ego boost, for sure, as it does wonders for a sense of professional worth. It also paints the clinician as the hero who fixes things where math falls short.

However, when faced with the threat of more powerful statistical insights, the psychologist's brain goes into high gear. With this, the doctor perhaps only remembers, or fabricates, times when her diagnoses were correct. She downplays the success of the tests. And so, we see that her earlier internal narrative survives, and perhaps even more strongly supports her view that she has unique and irreplaceable value. I would have hoped that, as a psychologist, she would have known better.

This might not be about her deliberately snubbing evidence. Rather, her interpreting brain likely dodged new information as it strove to maintain her comforting beliefs.

Grasping this quirk offers counselors an often-resisted pathway to professional growth. If we admit that we, like others, can stubbornly rely on the stories we tell about ourselves, we can begin to escape their grip.

We have seen that such errors are a widespread human trait, and possibly impediments to achieving accurate views of many things. As evolutionary biologist Stephen J. Gould[14] concluded, "the stereotype of a fully rational and objective 'scientific method,' with individual scientists as logical . . . robots, is self-serving mythology."

It is only by recognizing, and accepting, our common weakness of confabulation that we can advance to more accurate and useful self-understanding. But because our shortcomings are so difficult to see in ourselves, this can be an enormous challenge.

In the mind's deep haze,
Fiction blends with waking life—
Truth and tales amaze.

CHAPTER SEVEN

THE MYTHS AND MAKING OF THE SELF

Part I: Exploring the Elusive Nature of the Self

Now let's zoom in on what is likely a cornerstone of our most pervasive and unquestioned commonsense thinking: our overall concept of the human self.

Our common Western thinking suggests that this "self" often resembles an enduring person that resides in our skin from birth through death. This "self" includes an internal observer that watches the person we turn out to be during our lifetimes. Virtually all of us seem to sense this apparently observant entity. It's not a new concept. In his 1890 book *The Principles of Psychology*,[1] psychologist William James called this supposed internal observer "I."

We tend to associate this "I" with other concepts. For instance, you might consider yourself clever, patient, or a bit of a wiseass—maybe even a creative introvert. William James described these varying selves as "me." They are the *object* of what he thought the "I" observed. So when we talk about a sense of self, we're really discussing a blend of both "I" and "me."

Next, let's explore some questions that cast doubt on the traditional model of a self-observant "I." Is it in our head, our heart, or spread throughout our body? Could medical professionals pinpoint it? Nope. Further, examining brain function shows us that information doesn't go to just one centralized "command center" that's calling the shots. So much for the idea of a conductor in our head directing things.

And what about the consistency of this "I" as we age? Our bodies undergo drastic changes from infancy to old age. Our memories, beliefs, tastes, and even our needs shift over time. Yet many of us still believe that this supposed "I" has remained constant throughout these changes.

Further complicating matters, commonsense thinking often equates this "I" with our body or mind, though we can hardly articulate what the mind is. Is it just the brain, as Antonio thought, or is it more? Yet nearly one in four of us claims to have experienced our "I" momentarily leaving our body, perhaps floating near the ceiling in what some call "astral travel."

Moreover, common phrases like "my body" imply a separate "I" that owns the body. What's more, people who have had severe strokes or brain injuries often claim that their sense of "I" remains intact.

The Impact of Our Concepts

A near-constant inner dialogue—or voices with diverse concerns and agendas—constitutes what most of us call our thoughts.

These thoughts divide the world by breaking things into concepts and categories, each with a distinct name. We have thousands of these labels, which include every person we know (as well as the concept of person), every object and action, every duality (dead and alive, safe and unsafe, light and dark, useful and useless, and so forth), and everything else that we can imagine. This includes, of course, the model of "I." Each concept, category, and name is essentially a tiny model that helps us navigate the world.

Our senses help us categorize the world into these concepts. Whether visual, tactile, olfactory, or auditory, these natural sensory divisions are how part of our biology shapes our ideas. For instance, our idea of beauty often hinges on visual elements—say, a colorful sunset or a masterfully painted portrait.

We need these names, concepts, and categories to survive, to reason, to solve problems, and to communicate. But, as we have seen, naming or categorizing or explaining something, while useful, doesn't necessarily capture an ultimate truth.

For instance, the term *sunset* describes something real that happens every day, but the term itself—a relic from pre-Copernican days—is false. We know that the sun doesn't revolve around the Earth, and so it doesn't really "set." It's the Earth that is moving. But we say "sunset" because the term is useful, even though we know it's inaccurate.

Similarly, we conceive of a planet as an object that orbits a star—an object large enough to form a round shape by the force of its own gravity, yet not large enough to ignite like a star because of thermonuclear fusion.

Now imagine that, because of some unseen force, a planet near the star Proxima Centauri suddenly breaks free of its orbit. The moment it does, it ceases to be a planet, because a planet (by definition) orbits a star. But of course, the object itself did not suddenly disappear or morph into some utterly different form. It simply ceased to fit our conceptual definition of a planet.

In essence, *planet* is a concept—a model—like the idea of an "I." Nature does not make planets or "I's", as both are nothing but more of our mental models. Nature instead creates enormous rotating objects and human bodies that our left brains conceptualize and name. Planets and "I" are both useful ways to think about certain celestial objects and humans, but both are concepts of reality rather than reality itself.

Knowing this, we might wonder what would happen if tomorrow all human beings recognized that the model of an internal "I" was only a useful fiction. What in our everyday lives would change?

Perhaps surprisingly little. We would almost certainly continue to use personal pronouns such as you, me, her, ours, I, me, mine, and so on for convenience. After all, in the 21st century, we still speak of watching a sunset, even though we know that the Earth orbits the sun, or dialing our phones, even though they do not have dials.*

Where Is the "I"?

Now let's close this circle. We tend to accept many of the narratives our brain spins for us. As discussed earlier, this includes tales about who we are and a hypothetical "I" that supposedly hangs out somewhere within us, usually in the head. We often think that our internal dialogue reveals this "I"—an ever-present, invisible entity. This is like imagining that there's a person, a human within our bodies, generating the internal chatter we call thoughts.

Perhaps you already see the problem with this inference. It's like telling yourself that the sound of music is coming out of your smart phone, so the Boston Symphony Orchestra must be inside your phone. No investigation by

---

* Of course, we can easily update any story about ourselves as new information becomes available. Anna did this during her transformative journey. She greatly changed her sense of "I" when her hypnotherapy sessions revealed the transient and unstable foundation of some of her thinking. This is exciting because Anna showed us that neither her "I" nor her "self" were set in stone. She showed that new experiences can help us learn and grow in surprising ways.

a phone tech is going to locate a tiny orchestra inside your phone. Likewise, no study of the human brain has found a hidden "I" inside.

It's important to note that many people have rejected the idea of one, or even multiple, internal "I's." We will soon see that some Eastern philosophers, for example, do not even picture anything resembling a person inside our bodies governing us from infancy to death. They see this make-believe "I" as nothing more than a compelling and useful idea—a model—rather than an actual entity.

It would be hard to imagine a more unconvincing—or empirically unsupported—model than that of an internal, often enduring, self-observant "I."

The more we investigate the concept of a lasting observant "I" inside us, more or less separate from other things, the less sense it makes—and the more futile our search for this "I" seems. Eventually it becomes clear that this "I" that we assume, believe in, and appear to have is in fact nowhere to be found. It's just a model. Our eternal "I" is only a concept, or a process—and one that might shift over time.

Part 2: How Does the Idea of "Self" Happen?

Science doesn't have a clear single answer to unravel the mystery about what creates our sense of "self." There are, however, plenty of theories. As you might guess, the left-hemisphere interpreter (identified by Gazzaniga) might concoct stories about a "self," just as it does about why we select certain stockings or choose M&Ms. But let's look at this profound issue in a little more detail.

Inside versus Outside

First, our brains have a natural skill: They are hardwired to draw a line between a physical "inside" and a physical "outside." This borderline is important for our later creation of a coherent narrative of "self." It says, "Hey, this is me. I'm here to stay and I'm not part of that tree over there."

We need a complex brain to accomplish this feat. The neural networks that help with this are spread throughout much of our brains. But let's talk about just one of these areas called the anterior precuneus. It's near the top of the head and helps us develop a sense of a *physical* self.[2] It helps us tell, or create, a difference between a "self" and the outside world.

If it's damaged, you likely have much less of a sense of your physical "self." Thus you might also suffer from disruptions in knowledge about your location, movements, and the sensations that make up your physical being.

Let's think about why it's important for us, with the help of the anterior precuneus, to distinguish our physical bodies from everything else in the world. Any animal that could not tell itself apart from its surroundings would be helpless. What if you didn't know you're hungry because a raccoon took your food, or didn't realize that you need to run from the bear sprinting toward you? And so, for survival, telling the difference between what is "inside" and "outside" is essential.

Tickles and Self-Preservation

Let's talk tickles. Perhaps surprisingly, they give us insights into our brains' evolution. They showcase our ability to separate a "self" from everything else that's out there. This skill helps us not become snacks for animals. For instance, if a bug crawls up your leg, the tickling grabs your attention. It's clear that this is an alarm system because this kind of tickle turns our awareness toward threats such as biting or poisonous insects.

Have you ever noticed how you can't tickle yourself? That's because "gargalesis," another kind of tickling that makes you giggle, is almost impossible to do to yourself. This is because the brain can be good at telling the difference between actions originating in the brain from actions coming from the outside world. Our nervous system anticipates and recognizes our own movements, making self-tickling almost impossible.

In all, tickling displays some of our brains' many amazing abilities.[3] It shows how we've evolved to view ourselves as separate from the world, which is part of the foundation of our sense of "self."

Memory and Self-Narrative

Memory, too, plays a role in shaping our sense of "self." Recalling facts about ourselves—whether it's what we did in third grade or our favorite ice cream—contributes to our personal narrative. Our brains even treat these "autobiographical" memories about ourselves differently than they do memories of other events. Special neural networks in different parts of the brain are dedicated to managing this unique type of personal memory.

Another key player helping to create a "self" is the prefrontal cortex, found just behind our foreheads. This region is important for abstract thinking, decision making, planning, and controlling our actions. The prefrontal cortex becomes more active when people hear traits describing themselves as opposed to others. It helps combine such bits of information into a coherent, enduring sense of "self" that stands apart from the external world.

So, even though we haven't been able to pinpoint a singular "self" within the human brain, we have identified regions that help form this perception, as well as some of the reasons they do so. We can think of this process as the brain's way of organizing a flood of information into the understandable narrative of "self."

Social Blueprints of "Self"

It's easy to think that our sense of self is just the result of neural circuits. However, by looking more closely, we can see that society and culture also play roles in shaping our perception of our own "self."

Different societies mold unique conceptual frameworks for the "self." However, it's important to not oversimplify our thinking, even within societies, by assuming, for example, that the East thinks one way and the West another. That would be like saying that all North American food tastes like it came from McDonald's.

Despite this, cultural perspectives are sometimes simplified for research purposes by being called either individualistic or collectivistic.[4] These aren't just academic ideas. Instead, they are social models that show up as measurable differences in how brains work. In fact, individuals prompted to describe themselves in individualistic terms have shown less brain-scan activity in the prefrontal cortex compared to those describing themselves from a collectivistic angle.

In individualistic societies, the narrative surrounding the "self" drills down on the idea of standing alone, distinct from the crowd. This shapes a perception of self that is somewhat fixed. In individualistic societies, your likes and dislikes, personality traits, and the other parts of your identity are seen as less influenced by your friends, culture, and society.

On the other side, collectivistic societies encourage a more flexible take on the self. Here, "self" is tied to one's group environment. You are more like a social chameleon. Your sense of self morphs depending on such things as your current group of friends.

Some indigenous cultures around the world, for instance, abandon the idea of an individual "self" that is focused on material gain. Instead, they see humans as deeply intertwined with their surroundings. This might be nature, a family, or the entire community.

Another example comes from Africa, a place home to the philosophy of *ubuntu.* This belief system pictures our own lives as part of the lives of everyone around us. It stresses the importance of everyone's responsibilities to others. With this, it strongly supports collectivism.

Meanwhile, some thinking from the Middle East incorporates spiritual dimensions into ideas of the self. Inherent in those spiritual aspects is the belief that the true "self" and all of humankind are intrinsically connected to the divine.

It's worth being curious about how these individualistic and collectivistic societal templates play out in our everyday experiences. Chances are that everyone's sense of "self" *switches between individualistic and collectivistic viewpoints* to different degrees regardless of our society's inherent stance. Perhaps when we're alone reading a book, our "self" leans more toward the individualistic. However, during a family gathering, our "self" can morph into a more collectivistic version to match the values of the group we've joined.

The dance between our brain circuits and cultural upbringing is thus much more complex than we commonly understand. What we do know, however, is that our sense of "self" is a fluid construct, shaped not just by the neurons firing in our brains but also by culture.

These, and many other philosophies about the self, often don't stay put. They can travel around the world, even influencing how other cultures think. Eastern philosophies are now accepted in most modern Western cultures and are used in some respected, and researched, approaches to psychotherapy. Even executives in New York City sometimes practice focused meditation and examine their own sense of themselves before scrutinizing their spreadsheets.

So, let's all remember to be open to the many conceptions of "self" the world has to offer and see how they all can be valuable. The rest of the world might have a lot to teach us.

Flattering Rationalizations and "Self"

Once we see that we conjure up ideas of our own "self," we might wonder even more about why it takes a certain form. We perhaps hope that a picture of "self" would depend only on rational thought. While it's tempting to conclude this is so, let's be honest. We are far from rational creatures. In fact, many experts think that humans are Olympic-level experts at rationalizing, in this case making up plausible and self-serving descriptions of ourselves.[5] This means that we often explain or justify (or even confabulate!) who we are with answers that make us look good or seem true, even though we might be neither.

Rationalization is not just an occasional quirk but a cornerstone of the "self" we create for ourselves. Rather than the "self" we think is based on facts, it can be more of a flattering story. Perhaps this is like what happened when Anna created the seemingly helpful but ultimately false Sarah. Just as Anna created Sarah, others might write flattering but false self-narratives where they are ardent atheists, devout Christians, peacemakers, or people who thrive on adventure.

Moving on, many would wonder what determines why we keep some stories and discard others. One guess is simple: Our self-image is partly maintained by how we desperately want to be seen by the world. It's no big shock that we go to great lengths to ensure that our own sense of self is in harmony with this wish.

For example, think about some greedy figures in banking, industry, and technology. They might rationalize their nearly insatiable lifelong thirst for wealth by clinging to the belief that, deep down, they are good people, despite their questionable business ethics.

History gives plenty of examples of people using rationalization, without end, to present themselves as the good guys. Think back to the "Doctrine of Discovery." This rationalization justified the domination and oppression of Native Americans and other indigenous cultures under the cloak of divine destiny. According to this doctrine, land that was vacant or inhabited by non-Christians could be "discovered" and therefore claimed and ruled. It was a convenient way for colonizers to see themselves as noble heroes rather than as oppressors.

We see therefore that our continuing sense of who we are might partly be controlled by misleading rationalizations, which can serve a larger purpose. Although often comforting, these narratives can also mislead.

Anthropomorphism and "Self"

We humans are experts at anthropomorphism—the knack of attributing human traits to animals, objects, or even abstract concepts. It's second nature to us that we not only see smiling faces in clouds but also feel compelled to say "thank you" to our voice-activated devices.

In literature, this human-centric viewpoint shows up beautifully. Take Herman Melville's *Moby Dick*, for example. Melville transforms and elevates a whaling ship using humanlike qualities: "She was apparelled [*sic*] like any barbaric Ethiopian emperor, his neck heavy with pendants of polished ivory. She was a thing of trophies." Here, the ship is more than just an assembly of wood and nails; it has a personality and a sense of "self."

Just as we give a ship or a voice assistant a persona, that is, a role to play, we might do something similar with ourselves. We gather our experiences, thoughts, and emotions into a unified self that is seemingly permanent. This sense of "I" becomes a mental handle and helps us navigate life. But, once again, it's crucial to realize that this is more of a narrative construction than a solid entity.

So, as we anthropomorphize objects around us, let's also pause and consider that the "self" we seem to feel so deeply within may be just another narrative woven by our brains—a handy story that helps us navigate the world but that may not be tethered to hard facts. Our inner sense of "self" may be as real, or as fictional, as the human characteristics we so easily project onto the world around us.

"I" and Social Interplay

The idea of an "I" helps us seem to peer inward as an observer. This model uses the same lens we apply to observing our relationships with other people. It's as if we create an internal "I" as a companion to our encounters with others. For example, picture yourself deciding whether to walk across a frozen lake. "On the one hand, I know that the ice is very thick, so it's probably safe. On the other, I know that if I fell in, there would be no one nearby who could rescue me." This internal debate mirrors conversations between two people discussing what to do. It is easy to comprehend.

It's not surprising that this model of people perceiving themselves as "I," or even as multiple "I's," is deeply ingrained throughout almost all Western cultures and languages. It's like a shorthand that offers an easy way of talking about "oneself" that aligns with existing social relationships.

"I" and Self-Preservation

Some believe that the idea of even one interior "I" helps us in yet another way. Perhaps it gives us something beyond our physical bodies to look after, nurture, and keep safe from danger. Imagine having a friend inside your brain that you would want to protect in the same way that you would care for a child.

There is even some speculation by evolutionary psychologists that this sense of an internal "I" is an evolutionary strategy. It is one that might make us better caretakers of both our "selves" and our DNA.

Language Shapes "I"

Language is a key element in first building and then strengthening the model of "I." Once we begin to learn to separate out aspects of the world and label them conceptually through language, the process of building a sense of a separate "I" has begun.

Eventually this comes easily to us. Look, for instance, at how effortlessly we can switch from the internal concept of one "I" to the notion of multiple "I"s. Not only do we do this frequently, but we often don't even recognize that we're doing it.

For instance, when you say, "I caught myself obsessing over my hair," you've referenced two "I"s: one to obsess and another that caught the first "I" in the act.

Even compound words can imply multiple "I"s and further cement this misperception. Words such as self-confidence, self-esteem, and self-consciousness inherently create an internal sense of dual selves: one "I" to feel good (or anxious) and another "I" for the first "I" to feel good or anxious about.

"I" and the Mundane

Even something as commonplace as food can impact our sense of "I." At first glance, munching a burger or eating a salad might not seem to influence our self-identity. But food changes our mental state, our mood, thinking, and self-esteem. Especially during stressful times, going to a fancy restaurant can transform your sense of "I." It could even help you to feel like a different person.

The Multifaceted Nature of "Self"

We have seen that understanding the origins of our sense of "self" is far from straightforward. This identity isn't shaped by neurological processes alone, nor is it solely a product of our societal and cultural influences. Instead, it is a combination of things. These include biological, psychological, and cultural factors. Each contributes to the "self's" ever-changing form. To isolate one thing and say it is the one source of "self" would be like saying pizza is just cheese: an oversimplification!

To further grasp the intricacies of "self," think of it as a painting in progress, where you are both the artist and the canvas. You are coloring your own existence while also being shaped by your experience of seeing the work taking shape. This holistic view of identity resonates with certain Eastern, as well as with other, philosophies that envision interdependence and change.

Stretching this further, consider that even the social frameworks that help shape our identity are not constant. Thanks to such things as historical developments, economic shifts, and technological advances, they have also evolved. Just think about how differently you would think of yourself if you had been born 200 years ago or in a much different culture.

Becoming Attached to Fictional Selves

It's no surprise that people are often enamored with their sense of self. The irony, of course, is that this self is a creation, one that people often defend with their lives. Think about author (and amazing tattoo artist!) Brandon Garic Notch's wonderful quote, "As soon as you are born, you're given a name, a religion, a nationality and a race. You spend the rest of your life defining and defending a fictional identity." It powerfully highlights how becoming too attached to an identity can cause a lifetime of suffering.

In contrast, many people who follow certain Eastern traditions do not rigidly cling to their ideas about self. Instead, they embrace an ever-changing self while recognizing the impermanence inherent in all we experience. They highlight the importance of questioning, flexible mindsets that welcome new ideas as opportunities for growth.

CHAPTER EIGHT

THE LIMITS OF A RELATIVELY FIXED
SELF IN A DYNAMIC WORLD

Some contemporary psychoanalytic thinking, which has grown out of Freud's foundational concepts, challenges ideas of a somewhat fixed "I" and "me." Rather than envisioning a comparatively static and enduring self, it recognizes the dynamic, emerging nature of how we are thought to experience ourselves. In this model, *I* and *me* are verbs, or unfolding activities, rather than nouns. They refer to an ever-changing process of self-construction, rather than to an idea of something resembling a more fixed internal being.

It is partly from such insight that the valuable concept of "selving" has emerged. *Selving* is a term for the active process through which the self constantly transforms. It evolves with each new experience and insight in an ever-changing process. As one theorist puts it, "Our thinking, feeling, and acting are not what our self does; they are what our self is."[1]

Following this model, your sense of self is in a perpetual state of flux. One day you might find yourself to be resilient, and the next day vulnerability takes its place. Your "self" can include not just a single individual but an entire family or, as we saw with Anna, different personas within the same body. Anna's struggle with overeating showed how some of these shifts at times felt alien to her self-concept. Yet, through therapy, they soon converged into one cohesive yet constantly evolving "self."

Such shifts in perspective can hold immense therapeutic potential. This can occur in a person struggling with urges to overeat. At first, these desires might be mysterious intrusions—concepts with significant meaning that are foreign to the person's concept of self, or "me." These urges are sometimes frightening and seem insurmountable. However, through counseling, some patients come to see that they are not foreign invaders but impulses that stem directly from their own life experiences. This realization can make these urges seem less daunting and more manageable as they become integrated into someone's evolving "self."

Confronting Loss by Abandoning a Fixed-Self Model

Anna's struggles remind us of the troubles with fixed-self thinking. In her case, this view caused turmoil because she saw herself as separate from the world, doing battle with such things as other students and unrealistic community standards for body size. This might have been reflected in emotional eating. We also saw reflections of this thinking in the explanations of childhood torment that she concocted to explain her weight issues.

Anna began her journey with Antonio by viewing life through a static self-centric lens. At the start of her treatment, she saw herself as a long-suffering, lone individual, cleanly set apart from the world around her. She approached life as a battle against the problems that the world would bring. Certainly, she lived her life with a sense of a relatively unchanging "me" versus "everything else."

Let's first think about how Anna's life might have unfolded if she had continued to hold on to this rigid and self-centric perspective. What if she faced a totally unexpected loss, like the passing of her mother? In this scenario, she could have coped by turning to emotional eating. We can think of this as her trying to repair a fractured "self," trying to fill an emotional void by seeking temporary solace in food.

Now let's consider what would likely have happened if Anna had taken on board some of the wisdom she gained in her sessions with Antonio, that her ongoing sense of herself as a person inside of her skin was only a malleable idea. With this, she could have shifted to seeing herself as a constantly evolving amalgamation of emotions, thoughts, and experiences, and so she perhaps would have navigated her grief slightly differently.

This insight could have resulted in the recognition and appreciation of her mother's place in her life journey as being always a part of who she has become, rather than seeing her absence as a permanent void. She might even have wondered, "If I am not a permanent self, then who is it that is grieving?"

Although her grief would not have disappeared, this perspective perhaps would also have helped her to embrace life's natural impermanence. It can be easier to let something go and accept loss when you realize that, to begin, you never really had something permanent. With this, losses can become milestones of meaning rather than sources of endless suffering.

Anna's Journey: Leaving a Fixed-Self World Behind

Anna's experience shows how therapists can occasionally help patients through techniques that blur the concept of a relatively static "self." We saw that, when done correctly, this can lead to liberation and self-discovery.

Initially, Anna's therapy affirmed that she was a unique identity, distinct from, and coping with, external events. At this stage, she was simply Anna, coming to the therapist to help her deal with her eating habits.

However, through his action of encouraging Anna to summon up "Sarah"—an alternate internal "self"—Antonio introduced the idea that Anna was not a fixed entity but rather an ever-changing process composed of her experiences. He used hypnosis to help her to see herself from this different perspective. This viewpoint highlighted the fluid nature of her identity and was consistent with some Eastern thought.

Despite this new flexible sense of self, Anna was still tasked with shouldering her own transformation. This, of course, is a typical Western approach, once more relying on the concept that we all are the masters of our own lives.

As the therapy unfolded, we saw a profound therapeutic shift occur. This happened when Anna's struggles with food were seen as intertwined with the teasing she thought she had suffered.

Although she once thought that childhood ridicule had caused her weight gain, she soon realized that her story was a product of her own mind rather than of factual events. Sarah, an entirely new person within her skin, was also her mind's creation. Her insight showed her the human mind's power in writing self-stories that, while not reality based, greatly influenced her perception of herself. It was enlightening and liberating.

Thus, through Anna's story, we saw the profound impact of questioning and learning, understanding how such open-minded investigations can lead to growth and healing in unexpected ways.

When the Mirage of a Persistent "I" Misleads Psychology

Adopting the concept of a lasting "I" can hinder us when it comes to healing and self-discovery. This happens when it limits our openness to treatments that fall outside this model.

For instance, if we're wedded to the notion that our "I" will forever be plagued by past traumas, we could easily overlook true psychological solutions. Think about mental health patients who, for example, turn away from medications because they see them all as merely a futile way to avoid the inevitable.

Even scientists can think, sometimes productively, sometimes not, in terms of a stable internal "I." For instance, it's tempting to think of our brains as having an *all-knowing and independent* "inner CEO" or "secretary," a sort of central entity within us that makes decisions and filters information. This idea is sometimes linked to theories about the left-brain interpreter or the "executive control system" located in the prefrontal cortex, one of the most recently evolved parts of our brains. According to this model, this "control system" serves as a sort of manager, selecting what information gets passed to our conscious awareness, much like a secretary who screens calls for a company's director.

The allure of this theory is its simplicity. It's easy to assume that there's a smart, largely autonomous "I" inside us that makes decisions and shapes our awareness. This theory is even echoed in Anna's journey, where her "inner voice," Sarah, seemed to make independent decisions and offer selective insights about her overeating habits.

However, this way of thinking has pitfalls. For one, it can lead us to think that every bizarre or inexplicable action a person takes is the work of their internal "secretary" making peculiar decisions for reasons we don't understand. It's like attributing someone's actions to a created personality—like Anna's "Sarah"—and then trying to rationalize why an entity such as Sarah kept quiet all these years, fueling Anna's detrimental habits.

Just as Anna's inner voice turned out to be based on false memories, this supposed lasting inner "I," secretary, or CEO might also seem to mislead us. "It" can feed us incomplete or even deceptive information. This notion questions the reliability of any so-called hidden inner dialogue we perhaps think we have.

Moreover, the idea of an inner CEO can steer us away from evidence-based understanding. As we've seen, mental models and frameworks can be both illuminating and misleading. If an observer were to watch your life unfold from start to finish, they might buy into this CEO theory, assuming there's a stable entity directing your actions even as your body changes over time. But just as theories like bodily humors or demonic possession were once taken as legitimate explanations, the idea of a largely autonomous and consistent inner secretary or CEO might be more about storytelling than science.

Nonself and Nondual Awareness

Let's look at how an Eastern concept called "nondual awareness" melds beautifully with the idea of selving.

Nondual awareness includes the idea that everything is interconnected. It teaches that there are, in fact, no subject-object divisions and pictures the world more like a seamless web rather than isolated entities. This philosophy emphasizes that the categories that we use, like good and bad or self and other, are just mental sticky notes and are not an ultimate reality.

As we have seen, selving is about the ongoing, never-stagnant process of becoming who we are. It's like a river that's always flowing.

So, let's circle back to Anna, Antonio, and Sarah. They each seem like separate people exploring Anna's mind. But from the model of nondual awareness, they all make up only one all-encompassing consciousness. It's as if Anna's journey, including her hypnotherapy, isn't just made up of individual episodes. Instead, we can look at it as facets of one larger, more important, interpenetrated reality that includes even Antonio as part of Anna's experience.

Think about selving as if it were a kaleidoscope. Each piece of colored glass represents part of who you are. Every one of them continuously shifts to create new patterns as light passes through them. Now, nondual awareness is like the light that brightens the whole kaleidoscope. It doesn't ignore the significance of each little glass shard. Instead, it shows how they all make up the whole in a way that is essential to the existence of that whole, and far more beautiful than just the sum of their parts. After all, a tiny piece of colored glass isn't particularly significant in and of itself. But in its existence as an indivisible part of the whole kaleidoscope, it is a thing of beauty.

With this perspective, we can see that Anna's journey shows us that understanding ourselves isn't simple. Sometimes we find detours, like false memories and alternate selves. When this happens, it's important to not get discouraged. These detours often bring a better grasp of who we are.

Eliminating and Shaping the Self

Some people encourage others to let go of their sense of self. They might try to accomplish this partly by encouraging people to let go of desires, seen as persistent urges that often offer only fleeting joy.

We can look at giving up possessions through the psychoanalytic lens of selving. What exactly happens when we sell our Corvette or renounce a desire? According to selving, you aren't just parting with a luxury or a fleeting craving. You are instead letting go of pieces of your "self" as you previously understood it. You are going through more of an existential shift, perhaps like shedding a layer of skin or peeling away part of your identity.

Some Eastern approaches also encourage meditation to manage desires and emotions, as well as to quiet worrisome thoughts about the past and future. Instead, being immersed in the present moment can help fade the perception of a separate self and instead emphasize our connections with the world.

Changing Life's Script by Questioning Language

Let's take a look at how we can use language to expand our worldview and reshape our sense of who we are. An intriguing concept called "linguistic relativity" proposes that our language acts like a lens filtering and coloring how we see and understand things.[2] Think of this idea as being a bit like trying to express concepts like "love" using only emojis.

Jorge Luis Borges's story "Tlön, Uqbar, Orbis Tertius"[3] gives us a fascinating thought experiment. In it, he introduces us to a fictional language from the imaginary world of Tlön. This language throws out the usual grammatical rules. Nouns or even the idea of "self" or "I" are nowhere to be found.

Now, speaking in Tlönian, how would you describe watching the moon rising over a river without using nouns or the idea of self? You can't say *moon*, *rose*, *river*, or *I*. Borges tackled this by saying, "Upward, beyond the onstreaming it mooned."

He shifts focus away from the moon and self and focuses on the vibrant notion of upward motion. His phrase "beyond the onstreaming" also abandons the noun *river* and instead turns attention to the river's mesmerizing flowing quality, i.e., "onstreaming." Like this, his phrase "it mooned" focuses on an event rather than on someone, a "self," watching it as a physical object.

This thought experiment makes us question both our identities and our traditional sense of self. But it also mirrors at least one part of Anna's self-discovery: that language (such as Antonio's instructions) changed her sense of herself. There, she found that her grasp of reality was permeable. It partly relied on models suggested to her, such as the memories she invented about childhood teasing. And so, both Anna's insights and the language of Tlön further challenge how we think about and perceive ourselves.

We might wonder how the words we choose, as well as those we discard, shape our lives in other ways. We just saw that Borges created a world where certain ideas such as the concept of discrete noun-objects including "I" didn't exist because they had no corresponding words in language. With such a change, the notion of being an isolated person could neither be expressed nor imagined.

Following Borges's thinking, we can picture societies that would avoid words that emphasize division, conflict, and separation. Without language to express exclusion, perhaps problems like isolation, prejudice, and self-interest would become trickier to grasp and become somewhat less common.

Of course, we should be cautious about pruning language and ideas. We might therefore instead consider focusing on *adding* terms that celebrate our connections with others and the world. Perhaps adopting words such as the Nguni Bantu term *ubuntu* would be a step forward. As we saw, it roughly expresses the idea that "I am because we are." Using it might help show how our bonds with others help mold our identities. Ideally, this would not only nurture a deeper sense of our interconnected selves but also lead to more cooperation and harmony.

Overall, the misstep is not that we have encouraged a convenient, perhaps partly language-generated illusion by saying "self." Instead, the problem lies in forgetting that a stable "self," as conventionally understood, is an illusion. As we have started to realize—and as we will soon see even more clearly—when we take illusion for reality, we can quickly stumble into trouble.

CHAPTER NINE

THE PROMISES OF A FLUID SELF

Optimism and Change

There is something liberating about knowing that our concept of a self is not fixed but instead is a work in progress. Think about somebody going through a "midlife crisis." They perhaps find a new spouse, buy a fancy car, or take up unusual or dangerous hobbies. Maybe it's even a career change at 50, or a '58 Corvette at age 70. Some people look at these as problems.

However, we can also view these changes as signaling long-overdue awakenings. Each can spotlight the wonderful potential for growth. Anna, for example, recalled nonexistent traumas. Even though they were not genuine, they opened her mind and could have stimulated important change. She's not alone.

In the best cases, any of these changes can create a more dynamic sense of self by opening a universe of possibilities. People can see that they are never "stuck" and that they can always refine who they are.

Personal Reinvention and Therapy

Let's look at how this can play out in therapy. Acknowledging that the self is more like clay than stone gives opportunities for exploration and freedom. It can be a cornerstone of mental health care. Instead of the attitude of "That's how you have always been and always will be," a counselor could say, "Well, that's how you were in the past. Now, together, let's both work on helping you become the person you've always wanted to be." With this, therapy becomes an optimistic joint venture full of promise.

A more static sense of self can also stop us from accepting what we did in the past. Think about the last time you looked at an old photo and saw yourself wearing shabby clothes and ugly eyeglasses. You say, "What was I thinking?" But notice that this implies that you, *in your present state*, caused your past actions. But you are a much different person now and should not be held responsible for some things that you did in the past.

And so, adopting this flexible model of your "self" takes into consideration your constantly evolving identities. It is not judgmental. It gets you off the hook. You no longer brutalize your past self with your current wisdom.

Dire experiences can also trigger similar transformations in what is sometimes referred to as the "phoenix phenomenon."[1] This includes changes seen in wrongly incarcerated prisoners who later worked for judicial reform, and in soldiers who, instead of fighting, eventually advocated for diplomacy.

We can think of these positive changes not just as ways that unchanging people adapt to troubled times. Instead, we might see them as revealing deeper change in people. For instance, Viktor Frankl[2] had profound psychological insights and developed a new type of psychotherapy after his incarceration in a concentration camp. And so again we see that these changes don't quite appear as simple "adaptation" but as more of an improved and fundamentally changed self, although sadly through trauma.

From Belief to Betrayal

Thinking of the self as pliable can, on the negative side, highlight some of our vulnerabilities. For example, external events can greatly shape our identity in ways we cannot predict or, in many cases, even see as they occur. This is a playground for dictators and demagogues. They can, often as master manipulators, mold our "selves" to serve their own goals.

Imagine a fictional charismatic leader, Rajan Q. Patell, who has captured a small developing Asian nation's attention. He is a mastermind in turning people against each other. He knows how to convince them that their neighbors with different traditions and rituals are somehow bent on taking everything they value, even their freedom. In his propaganda, he highlights concrete, unimaginative thinking that discourages his followers from even picturing a better way. He is essentially a cult leader who so convincingly creates a frightening world, one full of conspiracy stories, that his followers abandon who they were earlier in life. And with his country's limited media freedom, his job is even easier.

His supporters, persuaded by his warnings and enamored by his eloquence and charisma, soon redefine their own identities to mirror his. Under his spell, they resort to violence, all the while convinced that their actions are for the good of the country. Even more frightening, they consider themselves to be "true defenders" of their land and culture at the same time they engage in destructive acts.

With this, we see how Mr. Patell has dramatically reshaped their self-concepts. Once changed, people still feel good about themselves, they think they maintain their moral values, and yet they become focused on destruction. By putting him in power, his morals have become their own.

Even many spiritually minded people in this community fall under Rajan Patell's influence; though well meaning, they end up distorting their own faith to fit his rhetoric. They cherry-pick from sacred texts to justify their violent actions—even those that run contrary to their faith's overriding positive teachings. Even doctrines that condemn violence are sometimes sidestepped. Instead, following their manipulative leader, they adopt a different "good self," one that sometimes condones destruction, rationalized as action for the greater good. And so, using the model of a flexible "self," those people, as well as their ideas of spirituality, are radically changed.

We have seen that the concept of a fluid self can, on the positive side, show possibilities for our growth, healing, and evolution beyond our past identities. Often, in people such as Anna, we can even see that the boundary between falsehoods and truth can, intriguingly, become blurred. Other times, however, viewing our lives through the lens of this model can clarify our susceptibility to manipulation and deception, even to the point of changing our own sense of ourselves.

We might wisely embrace this changing self, continuously explore new experiences, and avoid some common human vulnerabilities. We saw that Anna developed a new view of herself by discovering that she was far more changeable, pliant, and adaptable than she had thought. Her eating habits were not a fixed part of who she was but rather a part of an ever-evolving process of being—and for that reason it was susceptible to change. We can develop such broadened understandings of who we are, too, so long as we are willing to step into the unknown with an open heart, as well as the courage to accept the uncertainty and change that comes with it.

CHAPTER TEN

THE MYTH OF AN UNCONSTRUCTED REALITY

Shaping Our Own Worlds

Most of us experience our senses as straightforward, uncomplicated tools that give us fixed reflections of the world around us. We might, for example, think about vision as a process whereby input from our eyes is projected onto a kind of mental screen that an inner "I" observes, much like when we watch a movie.

However, the evidence reveals that our senses are far more dynamic than this commonsense notion. They often change as they continuously adapt to our environment.

In the mid-20th century, researcher Ivo Kohler[1] showed how our senses adapt in an experiment where he made goggles out of two colors of glass. Subjects who wore the glasses looked through blue glass when they shifted their gaze to the left and through yellow glass when they gazed to the right. At first, when subjects looked left, everything took on a blue tint, and when they looked to the right, everything looked yellow.

But after wearing the goggles for several weeks, people automatically adapted to the color distortions caused by the shaded lenses. Their brains compensated. Although they continued wearing the goggles, the blue and yellow tints vanished, and everything looked normal.

You can easily test your own ability to visually compensate by observing the same ripe banana under different lighting conditions.[2] Look at it in the bright sun, on a cloud-covered day, and at night under a full (or almost full) moon. It will always appear yellow, even though it reflects some very different wavelengths of light. As you will experience for yourself, your visual system automatically—and easily—compensates for these varying light conditions.

As you might conclude, this ability has profound survival value.

This occurs routinely not only in what we see but in what we imagine. For example, if someone who is thought to be in a hypnotic trance is told to look at a dandelion, they'll see it as yellow. But if the hypnotist then says that the dandelion is blue, many subjects will easily shift what they see from yellow to blue.

There are other dimensions of our visual experiences that show how we help create our perception of colors. Sure, colors are often triggered by light waves of different frequencies striking our visual systems. But there is nothing inherent in the light waves alone that would lead us to have sensations of a variety of colors. Instead, our *brains* generate these diverse experiences.[3]

In fact, with different brains or retinas, we would experience these light frequencies differently. For instance, some people who are colorblind "see" the same wavelengths that we perceive as red instead as gray. Because of the unique genetics possible in some women, they have what is referred to as "super vision." They possess four, rather than the normal three, types of cones in their retinas. This genetic condition, called "tetrachromacy," results in the production of far more colors than does the normal vision of most people.

Cultural Lenses on Perception

What's even more fascinating is that our perceptions are not just due to biology but can be shaped by culture. This includes optical illusions. You would

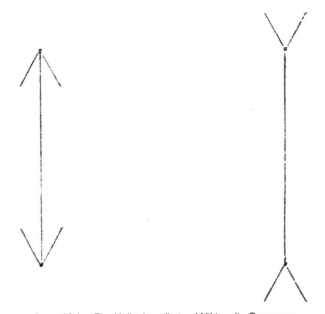

Figure 10.1. *The Müller-Lyer illusion.* Wikimedia Commons

think that most people would see them pretty much the same way. Not so. People from different cultures see them differently.

You have likely seen the famous Müller-Lyer illusion. It's the one that shows you two lines of the same length. One of the lines has arrowheads pointing outward, and the other has arrowheads pointing inward (see figure 10.1). This illusion tends to be especially tricky for Westerners, who often incorrectly decide that one line (the one with the arrowheads pointing inward, towards the center) is longer than the other.

Although there is some debate, some studies suggest that people coming from other cultures aren't as easily tricked.

How could this happen? We don't exactly know. Some people speculate, though, that people from Western cultures, with their focus on linear art and architecture, are "primed" to be fooled. However, people exposed to different architecture, such as that consisting of more organic shapes or natural landscapes, perhaps don't see visual objects in the same way.

And so we see that even our perceptions don't operate in cultural vacuums. This reminds us that our realities are not only idiosyncratic, but also ever-changing. What we see is therefore shaped by much more than just our eyeballs.

Beyond Passive Sensory Experience

Because of research on topics such as optical illusions, investigations such as Kohler's studies, and the work of many others, researchers now encourage us to think of perception as an adaptation-related *activity* rather than as a simple reflection of our surroundings. Put differently, from a scientific perspective, our perceptions are not passive reflections of the outside world. Instead, they are in part constructive and beneficial *behaviors* that *respond to* and help *shape* the world.

It's easy to think of toolmaking, speech, and communication as behaviors. It's not as intuitive to recognize that our perceptions also involve behavior.

Acting as if the color of an object is outside us, intrinsic to the object—rather than something partly created in our heads—is a mistaken idea. But it's also a useful way of imagining things. It helps us to select ripe fruit and heed stoplights.

Nevertheless, we need to disabuse ourselves of the idea that seeing color (or hearing sound, smelling odors, etc.) involves merely replicating something "out there" inside our head. Our behavior includes the partial *creation* of those sensations we experience.

Of course, we don't just create isolated sensations. At any moment, we integrate thousands of these separate sensations into our experience—a stream of ever-emerging events.

Let's pause to consider some implications. One commonsense idea is that our senses relay the external world to us—and, under normal conditions, do so more or less accurately. But we know that this is not the case. Indeed, our experiences blend what our senses relay and what our brains actively create.*

Now let's further expand our view. Since our world—our environment—is made up of our sensory and other experiences, this means that our brains partly shape our own world. To some extent, we each live in a unique world—one we've helped to create.

Think about Anna's weight-loss journey. Her invented memories of childhood teasing were not just like simple photos snapped of an "objective" reality. Instead, they were shaped by many factors, including her experiences, her emotions at the time, the therapist, and her beliefs about herself. Just like the people in the goggles experiment, she changed her world and made something new.

Knowing that perception is partly an adaptive behavior, rather than a simple reflection of an external world, feels unnatural. It's tough to accept that nothing outside us is replicated in our minds in its true form (if there is such a thing prior to perception!). And for many, it is almost inconceivable to think that the colors we perceive are partly mental *activities* rather than stable properties of the objects around us.

Color, and other sensory information, thus are adapted to our changing situations. In this sense, our perception of color—what we might call *color behavior*—is therefore like a baseball player's swings for each new pitch. Each new swing is different for, and responsive to, each pitch, not a distortion of the swing that preceded it.

This idea parallels the previously discussed concept of selving, where people actively create—and continuously recreate—their ever-evolving sense

---

* We also now know that our memories operate in a similar fashion. When we remember an incident, we don't just retrieve it, like pulling a file from a folder. In part, we actively create the memory. This is part of the reason why our memory of a single past event may change over time—and why two people can share the same experience and remember it quite differently.

of self. Such dynamic processes are more examples of how we both create as well as discover reality.

Insights from Science and Eastern Philosophy

Now let's examine the parallels between what we have examined so far and concepts from some Eastern philosophies.*

Much Eastern philosophy rejects the commonsense view of the world as consisting largely of stable, separate objects. Instead, it invites us to see the world as a whole, comprised of constantly changing and interacting events. From this perspective, the world is treated as a complex web of dynamic interconnections. Any perceived separateness between things (or between us and something else, or between us and the world) is an illusion.

This is consistent with daily life in some parts of the East where, perhaps partly due to increased population densities, people tend to see themselves first and foremost as part of (and therefore responsible to) a larger society. In contrast, here in the West, we are more likely to see ourselves mostly as individuals whose primary responsibility is to ourselves.†

Thus, when a Westerner looks at a rainbow, they might think that a rainbow exists "out there" and has many colors as some of its attributes. They can think that they then take in and experience these attributes through their eyes. They perhaps also imagine that the rainbow's colors inherently exist in the sky whether or not they look at it. But that's not what color is.

As Eastern metaphysics tells us—and modern laboratory research confirms—we, the rainbow, and its colors do not exist as separate entities. In the moment in which we experience the sight of a rainbow, what occurs is a web of relationships in which any separation is illusory. There is no separate object, no separate observer, no separate colors, and no separate anything else. There are no colors, smells, sounds, textures, etc. independent of a person or other creature who perceives—i.e., helps to produce—those perceptions.

The modern philosopher Sheng-yen[4] put it this way: "Sense organs and objects are spuriously named, because they refer to that which does not really

---

* There are of course many different Eastern philosophies. Here I use this general term to refer to Daoism, most forms of Buddhism, the Advaita form of Hinduism, and certain aspects of Confucianism.

† I am of course using the broadest of brushes here. Exceptions to (and variations of) these generalities abound.

exist. . . . Anything that exists through the coming together of causes has no true existence. It arises in coming together, and it perishes in separation."

With this, he is suggesting that the way people often talk about objects and their properties wrongly implies that they exist separately from each other, and from us as observers.

In short, the rainbow is an *interaction* that includes water droplets in the air, the position of the sun, the person looking at the sky, and the processes inside that person's head. But all of this is also part of a larger interaction that includes the atmosphere, the evaporation of moisture into the atmosphere, the winds that blew the moisture into the area, the blazing explosions on the sun, and so on and on, until—as perhaps you are beginning to see—the interaction involves the entire planet Earth, our solar system, and, ultimately, all of space and time. Or, as some philosophers might say, the rainbow occurs both everywhere and in no one particular place.

Quantum Entanglement and Nondualistic Thought

Many Eastern philosophies have embraced such "nondualistic" models of the world for millennia. As we would probably expect, they discard common boundaries between oneself and everything else. They do not, for example, think of the world using the dichotomies of observers and objects to be observed.

Eighteenth-century philosopher David Hume,[5] as well as others, such as some Native American thinkers,[6] have used similar models. Intriguingly, these beliefs have found a type of confirmation by at least some versions of the phenomenon of quantum entanglement. Here is an example of this kind of confirmation.

Visualize a pair of photons traveling billions of miles apart at the speed of light. When the spin of one is observed, the spin of the other *simultaneously* matches the first, as if the two were conjoined.

Even if the first photon could somehow signal the other, that signal would be limited by the speed of light. Yet the amount of time that elapses between the change in one photon and the change in the other is zero. No signaling seems to be taking place. Instead, the two photons, first pictured as separate things, don't seem so separate after all, something that defies a traditional understanding of separation and causation.

I'm not sure if anyone knows exactly how this phenomenon, called "quantum entanglement," happens. However, it is consistent with some nondualistic conceptions, suggesting that the apparently separate photons are instead somehow connected in a manner that is not obvious to us.

Some Eastern philosophies embrace the closely related idea of "dependent origination." This theory posits that nothing occurs independently—all phenomena occur only in relation to other phenomena. Simply put, nothing exists in isolation, and everything is intimately connected. You perhaps think this is only ancient wisdom. But even modern physics can echo these ancient beliefs.

The Interconnected Fabric of Existence

Here's another simple example of our sometimes convenient, but ultimately illusory, models of separateness. Imagine that you are asked to solve a complex math problem. You're sitting at a desk on which you have a pencil, some paper, and a calculator. After half an hour, using all the tools at your disposal—and lots of thinking—you arrive at the solution. According to our commonsense view of things, we would say that you, meaning the "you" that is supposedly inside your skin, solved the problem.

But getting to the solution needed more than just you. It also required the calculator, the pencil, the sheets of paper, the desk, and the chair you sat on. And for the paper to exist, there needed to be a tree, a paper mill, the earth in which the tree grew, the person and the chainsaw that cut down the tree, the truck that shipped the paper to the office supply store, the building material out of which the store was constructed, the bank that loaned the owner the money to start up the business, and so on and on and on. The math problem was not solved only in your head. *Instead, its solution came from a vast network of activity and interactivity that make up a larger "cognitive ecosystem."*

By this account, our minds are not isolated. Instead, they are inseparable from our physical, social, and technological worlds. The external world is not merely a backdrop for thinking. It helps make our mental experiences. Thus, in my opinion, the mind can often best be seen as a process that goes on far beyond us as isolated individuals.

CHAPTER ELEVEN

THE MYTH THAT WE CAN PEER INSIDE OUR OWN HEADS

M ost of us take it for granted that we can observe our own thoughts. When we do this, we say that we are "being introspective" or "looking inside ourselves."

Certainly, this is what we *seem* to be doing. A thought arises and we observe it. What could be simpler?

Yet this is like thinking that we "take in" colors with our eyes—except that the object of our attention is supposedly in our head rather than, say, in our backyard.

As we will soon see, however, reality is different from our widely held notions of introspection.

The Introspection Illusion

When we use this model—and think in this way—we act as if we had sensory organs in our brain. If this perspective were correct, there would perhaps be tiny biological cameras and microphones enabling us to monitor our thoughts, desires, impulses, and other things that arise in our heads.

Of course, neurobiologists have found little credible evidence of mechanisms for observing our own brains in this manner. Our brains do not even have their own pain receptors. If a brain surgeon were to slice into your brain without touching your skin or skull, you would feel no pain.

But there are bigger problems with this model.

First, there is the splitting of "I" into two: the "I" who has the thought and the "I" who observes the thought. As we saw earlier, this multiplication of selves is itself a flawed concept.[1]

Second, there is the notion that we "take in" thoughts in the same way that we "take in" colors. But, as we've seen, we don't "take in" colors that are seemingly "out there." Instead, we help create colors based on multiple factors, including our own biology. So when we help create thoughts or colors, we are not, in one sense, watching anything. In a sense, *when we assume that we are observing our thoughts, we are instead having a thought that we call an observation.* Knowing this means to reject the duality of an internal observer watching a screen inside our minds.

Thoughts as Behaviors, Not Experiences

To better understand what happens when we embrace the commonsense idea that we observe our own thoughts, imagine what would happen if two people—let's call them Maggie and Morgan—simultaneously try to learn about thinking but from different perspectives. Maggie is asked to "observe" her own thinking internally, with her eyes closed, while Morgan is told to look at a computer monitor showing a real-time scan of Maggie's brain. Would these two people experience the same things?

Clearly, no. Maggie has, *but does not witness,* sensations such as smells and sounds and emotions such as fear and relief. In a sense, she "observes" nothing. Meanwhile, Morgan sees a multicolored display of chemical and electrical signals. Their perspectives are, undeniably, distinct.

Morgan would conclude that he had seen "thinking," as revealed by chemical or electrical events. But Maggie would insist that she had *observed* her thoughts, sensations, and emotions, somewhat as if they belonged to a second person inside her body. This mistaken conclusion would have been like "Sarah's" belief that she could watch Anna's distress over her weight, i.e., like one person watching another who happens to reside in the same skull.

The Paradox of Self-Perception

So, Maggie and Morgan did not observe the same thing from different perspectives. In fact, Maggie *did not "observe herself" at all.* Indeed, since there is no internal "I"—no mini-me or autonomous CEO in the human brain—then no one ever *observes* their own thoughts. From this perspective,

---

* B. F. Skinner discussed our inability to observe our own thoughts when he wrote, "A severe limitation is to be seen in the organs a person uses in observing himself. After all, what are the anatomy and physiology of the inner eye?" Many contemporary psychologists agree. Steven Pinker, for instance, echoes Skinner's thinking when he observed, "The faculty with which we ponder the world has no ability to peer inside itself or our other faculties to see what makes them tick."

thoughts are not *experienced*; they are instead more like *behaviors that we perform.*

In other words, the idea that we can't observe our own thoughts refers to the fact that any thought we have about observing a thought is, well, just another thought. Informally, we perhaps call it an experience, as if there were a separate internal person watching a hidden television screen. But that's just shorthand.

It's counterintuitive to think that others can observe our thoughts, *but we cannot.* Nevertheless, this is one way that our mental behavior can be imagined. This is like having thoughts without either a thinker generating them or an observer watching them. In fact, more generally it has been suggested that our understanding of how our minds work might not originally come from "observing" our own thoughts and feelings, but rather from basic ideas we learn from our society.[2]

Dispelling a Dualistic Perception of Thought

We have seen that people often perceive a changing "self" as something that, nevertheless, somehow still includes an enduring and fairly consistent "I." Yet, we know that no such "I" can be found, let alone one that is relatively static.

Our experience of being both the thinker of our thoughts and the observer of them creates a parallel. We believe that we are somehow both of these things at once. Yet both are imaginary. *Neither the hidden thinker nor the hidden observer in our head exists in the way that our brain tells us it does.* This combination thinker and observer is like a demon, a bodily humor, or a breeze blowing through a hypnotist's windowless office.

One challenge, however, is that seeing through this illusion is not like knowing the secret to a magician's trick. Once you know how the trick is done, the magician's illusion is gone in an instant.

But it is far more difficult to get past the idea of a hidden "I" that is aware of everything we do. This idea is ingrained. It is almost always with us and is a part of our daily lives. It's not surprising that it's a tough habit to shake and, for practical purposes, there is usually little reason to do so.

Many of us can gain a fuller understanding of people by learning to see ourselves in this less fragmented way. Here, our thoughts are part of *us*, not something we observe from afar.

Letting Go of Mental Treasure Hunts

As an aspiring psychologist, I was often encouraged to both monitor my own thoughts and to help others do the same. Sometimes this philosophy helped. For instance, in a counseling session, I occasionally asked patients to "watch" their thoughts. By doing this mental exercise, some of them could better catch and diffuse those early signs of rising anger before they exploded into outbursts. This model sometimes increased awareness and kept emotions in check.

But at other times this approach fell flat. Digging for hidden mental gems wasn't worth it. In fact, it could backfire, especially for people prone to overthinking.

For these patients, I sometimes took the opposite approach. I encouraged them to ease up a bit and to be less fixated on their mental activities. This was, of course, a tall order for some.

But it did offer relief to some people who were able to spend less time ruminating. When they reduced their inward focus and kept their thinking in the present, they could often catch a break. No longer embroiled in a never-ending mental scavenger hunt, they could let their thoughts come and go rather than obsessing over their significance. What a relief.

These people no longer found themselves so wrapped up in their thoughts, the past, or the future. This practice helped them bypass judgment and fear. Some also describe it as involving slowly building more and subtler awareness of one's body and other phenomena. They thus gained a broader understanding of their lives, perhaps a little like appreciating the vastness of the entire night sky rather than focusing a telescope on one solitary star.

In all, we are greatly challenged by nonintuitive ideas about ourselves, including the notion that we can glimpse our own thoughts. This struggle reminds us that achieving greater self-knowledge is far from straightforward. It is instead a long path, scattered with both insights and obstacles.

CHAPTER TWELVE

THE MYTH THAT HYPNOSIS IS A UNIQUE STATE OF MIND

One commonsense model of hypnosis is that it involves unique mental processes that do not occur in our everyday lives. Neuroscientists, however, haven't definitively identified any specific physical events that occur in our brains *only* during hypnosis. In fact, much of the evidence shows that what happens in the human brain during hypnosis also happens (or can happen) at other times. Indeed, today the prevailing view, among credible theorists, is that hypnosis does not produce a unique state or mechanism that is not accessible through other means.

I'm not suggesting that mental processes in the brain don't change in people after they experience "hypnotic inductions." Often they do. But these changes have not been convincingly shown to be unique to "hypnotized" people.

Hypnotic Induction, Imagination, and Perception

To explore this, let's delve into the work of psychologist Stephen M. Kosslyn and his colleagues,[1] who have investigated color perception among people who were reportedly in "hypnotic trances." These researchers showed people, some of whom were hypnotized and some of whom were not, both gray and colored designs. Sometimes the researchers instructed the subjects to see the designs exactly as they were. At other times, they instructed subjects to mentally add color to the gray designs, or to mentally *remove* the color from the colored designs and see them in shades of gray.

What they found was fascinating. Brain blood flow in "hypnotized" people who were shown gray designs and instructed to see them in color mimicked the blood flow of nonhypnotized people who had been shown colored designs. This points to the flexibility of our brain's perceptual hardware, which can act similarly whether we're "hypnotized" or not.

Likewise, the brain blood flows of "hypnotized" people who were shown colored designs but were instructed to visualize them as gray resembled those of nonhypnotized people who looked at gray designs.

This suggests that our brains can act the same way when we look at something blue as they do when we are told, under "hypnosis," to see it as blue. In other words, *the brains of "hypnotized" people adopted a model of the world in which perceptions were based on instructions rather than on light wavelengths.* This probably does not surprise you by now.*

In fact, in our everyday lives we often perceive based primarily or entirely on instructions. Under the sharp tongues of drill sergeants, members of the military in basic training routinely perceive less pain when their muscles grow tired. Athletes generally perform better when they are coached—or when their friends shout, "You can do it!" Many of us "push past our limits" when we are told to ignore those limits and follow instructions instead. Ouija board players comply after they are told to perceive the planchet moving on its own. Some of us who are encouraged to hear divine voices seem to hear them as well.

Karl Scheibe and his colleagues[2] showed in another ingenious way that hypnosis is not a unique mental state. In this experiment, each subject was told that they would become deaf after being "put into a trance." After their "trance inductions," all the participants confirmed that they were, in fact, unable to hear. Then each was asked to put on headphones and read some text aloud. As they read, their own voice was played live over those headphones, but with a slight delay. This delayed feedback would of course have had no impact on people who were genuinely deaf. But it made all the "hypnotized" people with normal hearing stutter. Nevertheless, these folks all *believed* they were deaf.

Sociocognitive Models of Hypnosis

Virtually all psychologists, especially those who work with what are called *sociocognitive* models, agree that there is nothing necessarily unique about the mental states of people who are said to be hypnotized. Most of these psychologists see "hypnotic" behavior as another form of social behavior. Part

---

* Stephen M. Kosslyn's work is very important. It helps us ignore the traditional divisions between what we might think of as "normal" and "hypnotic" states. Therefore, we see that these distinctions are likely conceptual rather than factual.

of this is the belief that our culture provides us with some specific rules for determining when someone is hypnotized or not.

We are left with the question, *Who gets to decide whether someone is truly hypnotized?* When the hypnotist says they are? When the subject says so? (Recall Jean-Martin Charcot's patients, who experienced and/or acted out nonexistent physical symptoms. Many of them were certain that they had a unique mental state and that their uteruses truly wandered.) What if the hypnotist and their subject disagree? Whose viewpoint is correct? Who is the final authority? What if a subject is convinced that they were in a "hypnotic trance" an hour ago but a week later no longer feels that way?

Clearly, evidence suggests that, like "hysteria" or "wandering uteruses," what we call hypnosis—at least as most people usually think of it, as a separate state inside the brain—is a fabrication.

All of this could have interesting implications for understanding ourselves. It might mean that what and how we are during hypnosis is what and how we are *much—if not all—of the time.*

Seeing beyond this apparently false distinction—between a supposedly hypnotized state and our ordinary mental activity—requires a shift in our perspective. But we have all made many such shifts before—both individually and collectively.

For example, in his 1859 book *On the Origin of Species*,[3] Charles Darwin argued that dividing animals into separate species (in a manner similar to how we divide our experience into separate states of mind) was artificial and got in the way of deeply understanding our fellow creatures. Darwin encouraged us not to lose sight of animals' development, their relationships to one another, and the important role that environment plays in their lives. Darwin's insights engendered a major shift in perspective among most of humanity—a shift that profoundly informs the interconnected manner in which most knowledgeable human beings see themselves today.

Cultural Context and the Hypnotic Experience

Let me suggest that our method of determining whether someone is (or is not) hypnotized relies heavily on our *frame of reference*: a combination of our present perceptions, our past experiences, our knowledge of the world, our beliefs, our current emotions, and a welter of other information. Each person's unique frame of reference influences how they describe and think about the world at any given moment. This all occurs within larger frames of reference created by our culture, authority figures, media, and a variety of other influences and influencers.

A bear's frame of reference is that it's normal to hibernate, so a family of bears would be puzzled as to why humans don't curl up and sleep through the winter. Similarly, avid gamers might wonder why other people don't immerse themselves in computers and gaming conferences.

Our frames of reference determine not only what models of the world we use but which behaviors we believe show abnormal, as in "Arianna is in a hypnotic trance right now," versus normal, "Arianna is not hypnotized at the moment," behavior.

In chapter 1, Anna saw herself walking down a flight of stairs to a comfortable place. If you and I had been in the same room with Anna, we would have seen her sitting still in a chair with her eyes closed. How would we explain this mismatch between what Anna experienced and what we saw with our own eyes?

We might say, "Anna was hypnotized." But this doesn't mean we recognized or determined anything unique about her behavior. It simply means that we grabbed a handy, convincing-sounding explanation—one that was consistent with our own existent frame of reference.

Hypnosis and Our Need for Alternate Realities

There is an often-unacknowledged benefit to imagining that hypnosis is a separate state of mind.

Most of us like to think of ourselves (and each other) as reasonable, careful, and rational, at least most of the time. Few of us would describe ourselves as inherently impulsive, irrational, careless, and unreasonable.

Yet each of us sometimes *is* impulsive, irrational, careless, or unreasonable. How can we resolve this contradiction?

Many of us seem to do it by adopting a model of multiple states of mind. We imagine that in our default mode—our "normal" state of mind—we are reasonable, careful, and rational.

When we are "hypnotized," however, we sometimes think we act otherwise—strangely, impulsively, or irrationally. Under "hypnosis," we believe things that we would not otherwise believe. We declare that this "hypnotized," uncommon, and supposedly abnormal state of mind is responsible for our unusual behavior. By believing this, people compartmentalize to explain away such contradictions in thinking. We imagine, "Hypnosis made me do it."

This is fundamentally no different from saying, "The Devil made me do it." Or "I was possessed by a spirit, which made me do it." Or "My suppressed, unconscious anger at my brother made me do it." In all cases, it amounts to a denial of our common human frailties.

Such fabrications can help us maintain vain and comforting beliefs about ourselves by explaining that certain odd or embarrassing or harmful things we do occur only under highly specific and unusual circumstances—circumstances that we often proclaim are beyond our control.

This ability to create a variety of other supposedly unique, "altered" mental states can be seen as a tactic to help justify a variety of questionable behaviors. That said, this skill can be important to our well-being.

For example, picture you and me walking together in a park. Suddenly I turn to you and say, "My dead grandfather likes your hat. So do I." Would you wonder about my sanity?

Now imagine that you have been told by several people you trust that I can channel the spirits of my ancestors. Soon after we sit down on a bench to rest, I say, "I'm going to commune with my dead grandfather for a bit." Then I close my eyes and sit in silence for half a minute. When I open my eyes, I look you up and down, smile, and say, "My grandfather loved hats like yours. Even though he's been gone for almost a decade, I can still sense his spirit inside me. Looking at your hat, I can feel his pleasure."

Now would you wonder about my sanity—or would you just consider me eccentric, perhaps even charmingly so? Your sense that I am in a unique mental state, such as being in touch with my grandfather's spirit (even if it's nothing more than my fond memories of him), makes my message more credible. It also provides an incentive for me to regularly enter—or pretend to enter—that state or reflection.

Exhibiting (or appearing to exhibit) altered states of mind has other benefits. For example, it can create symbolic group markers, defining what group you belong to and to whom you give your loyalty. As Richerson and Boyd[4] observe, "once reliable symbolic markers exist, [natural] selection will favor the psychological propensity to imitate and interact selectively with individuals who share the same symbolic markers." In certain groups and contexts, demonstrating a presumed altered state or displaying a symbolic marker might help you belong—and therefore survive, thrive, and pass on your genes.

There is no more vivid demonstration of how presumed altered states serve as symbolic markers than Singapore's Hindu Festival of Thaipusam. During this two-day event, devotees pray, fast, and listen to chanted mantras until many of them exhibit trancelike behaviors.

Rainer Krack[5] describes a man at this festival who began to shake, as if losing control of his body. Then he pushed a metal skewer through his tongue while appearing to experience no pain. Other devotees attached fruit to fishhooks, then pierced his back with those hooks. The man then walked miles down the road wearing shoes with nails projecting up into his feet.

We could say that this man's pain control (or pain desensitization) was the result of a "hypnotic induction," or a chanting-induced trance, or an ecstatic religious experience, or any of half a dozen other explanations. Which one is correct? Which explanation feels most comfortable to you? If the man with the skewer were to give you an explanation, would you accept it as the truth, or would you seek alternative explanations?

The Festival of Thaipusam's drama gives participants many opportunities to display their connection with, and devotion to, their spiritual traditions and communities. Like people thought to be in hypnotic trances, participants are rewarded for their seemingly altered states of mind. Like many shows of patriotism and healing power, taking part in this festival affirms one's dedication to their group. It also affirms the power of potentially healing religious practices.

Clearly, being able to follow and reflect a group's beliefs or identity can be a powerful tool for both self and group preservation. This is one reason that gang members wear their gang's colors, why some Pentecostal Christians handle venomous snakes, and why hypnotists accept credit for psychological cures. So, while the dramatic behavior during the Hindu festival might seem unparalleled in much of the Western world, it shares elements with commonplace Western activities. Once again, we see that, as Steven Pinker observed, "our brains were shaped for fitness, not for truth."[6]

Navigating the Illusion of Control

Here's another common misconception about the "hypnotic state": It is sometimes thought to reduce or remove our free will.

Imagine that you drive to a nearby grocery store, buy a loaf of bread, and drive home. You would almost certainly say that everything you did on the trip was voluntary.*

Now, what if on the way home, a few blocks from the store, you take a wrong turn? Was taking that wrong turn voluntary? You would probably say no, because that wasn't what you wanted or intended to do.

So, what made you conclude that your actions on the way to the store were 100% voluntary and those on the way back were partly involuntary?

Psychologists Irving Kirsch and Steven J. Lynn[7] have examined how people determine whether their behavior is voluntary or involuntary. They found that the key difference hinges on the *context* in which an action occurs. When things go as we hope and plan—for instance, we drive to a store, buy bread, and drive home—we are likely to interpret our actions as voluntary and the result of our own free will. But when things don't go as planned—or when we do something that we would not consider fair, wholesome, or reasonable—then we are likely to say that it was involuntary. We adapt our rationale—our brain's story about what happened and why—based on the context.

Kirsch and Lynn's findings shed light on "hypnotic trances." These experiences allow us to credit (or blame) our thoughts, actions, and words on something other than our normal mental and emotional processes. Saying, "I would not have done that if I hadn't been hypnotized," can, in some contexts, even absolve us of guilt.

Kirsch and Lynn insightfully conclude, "Hypnosis is special, not because it involves any unique mechanisms, but rather because it illuminates normal human propensities that are not as readily apparent in most non-hypnotic situations." Hence, the only thing unique about hypnosis appears to be the name we give it.

---

* Kirsch and Lynn (1997) have challenged these commonsense interpretations about what guides our actions. First, they point out that humans can do all kinds of seemingly voluntary things based on rules they do not even know. Speech is a common example. When we talk, we almost effortlessly follow a welter of difficult rules. Yet most of us cannot explain many of these rules. (For example, try to explain why it is grammatically correct to say "fresh roast coffee" or "freshly roasted coffee" or "fresh roasted coffee," but not "freshly roast coffee.") We deliberately do many other highly complicated things, such as driving and playing sports, with little or no awareness of how we do them or many of the rules that govern them. Much of what can be thought of as our voluntary actions are at least partly controlled by involuntary, automatic processes.

Having this insight is not just a mental exercise. Instead, it pushes us toward greater self-knowledge. For instance, it shows that the distinction between hypnotic and nonhypnotic states, like the concept of self, is arbitrary. It also emphasizes the connections between our social environments and our self-perceptions, leading to a more profound understanding of our connectedness with our surroundings. Thus, we are digging deeper, like Anna did, to see new layers of human complexity and reveal avenues ripe for personal growth.

Hypnosis and an Expansive Self

One more point deserves some unpacking. We saw earlier that what we call a rainbow is, like changing our sense of self, a dynamic process—one that involves the entire universe.

Some recent, "relational" models of hypnosis posit a somewhat similar dynamic process. This process is not seen as only unfolding inside an individual's skull but as existing interactively—at a minimum within the interaction between the hypnotist and the client.

Some experts who embrace these models believe that a hypnotic subject can, through suggestion, essentially integrate the hypnotist into their own concept of "self." Others would view the same phenomenon as the "mind" losing its boundaries and seemingly merging with the mind of the hypnotist. In this view, a previously separate hypnotist and patient can assume qualities of having a single self or mind. In this scenario, the patient can experience hypnotic instructions as coming from within, not from another person.[8]

This model explains why, for example, a hypnotized person might quickly become more confident or optimistic by internalizing a hypnotist's optimism or confidence as their own. They would see these beliefs as coming from within themselves.

With these insights, it again becomes clear that the boundary between hypnosis and our normal states of mind is not distinct. Both are dynamic, relational processes where our sense of self can integrate external influences. Whether we are hypnotized or not, our brains are always interacting with rest of the world, blurring the lines between internal and external realities. This understanding once again suggests that hypnosis isn't a unique state but rather a natural extension of how our minds operate in everyday life.

CHAPTER THIRTEEN

HYPNOSIS AND OUR
INTERCONNECTED REALITY

I n the concluding sections of this chapter, I focus on seeing ourselves as part of one interconnected reality. To guide us down this path together, I use hypnosis metaphorically, comparing it to both commonsense, and some traditionally Eastern, ideas of "self." With this, perhaps you too can reimagine yourself as indivisible from all other facets of existence. *By introducing you to this thinking, I'm hoping that together we can transcend the limitations of commonsense knowledge and contemplate a new way of seeing ourselves within the world.* To get there, let's consider four basic mindsets.

I. Abandon Aspirations for Tangible Benefits

Traditional Western models sometimes, but certainly not always, view hypnosis as being mostly a reflection of people's thinking. This concept echoes Western philosophical and psychological traditions that have focused on the utilitarian value of scientifically examining people's thoughts.

For instance, you'll recall that Freud believed in an unconscious mind, where people hid their undesirable thoughts both from themselves and from others. Early on, he promoted hypnosis as one method to offer glimpses into this unconscious realm. He believed that hypnosis, used this way, would yield tangible benefits. For example, Freud thought that this technique would help his patients uncover past trauma, express the emotions associated with that trauma, and thereby alleviate their distress.

Freud's concepts focused on what went on inside people's brains. Research based on such models thus involves examining individuals' problem solving, reasoning, memory, and perception. All these processes are primarily practical and functional—so discourse around them tends to involve utilitarian intentions rather than abstract reflection. With his and other pragmatic approaches, hypnosis becomes an avenue to reduce pain, lose weight, decrease anxiety, or achieve some other tangible goal.[1] Such models almost invariably also posit (often by inference or implication) that each human being is a separate individual trying to cope with an outside world.

In contrast, as we know, some Eastern philosophical approaches, although diverse, focus on how the body and the "outside" world are interconnected parts of one seamless reality. This attitude is not, of course, primarily meant to lead to material or psychological rewards but instead to nurture a deeper understanding of ourselves within the world.

I see my own approach to hypnosis as a departure from many established Western frameworks. It is inspired by some Eastern ways of thinking, specifically in that it can favor more abstract and contemplative modes of thought rather than realizing immediate practical outcomes such as anxiety reduction or pain control. In fact, my perspective about hypnosis is less about day-to-day utility and more of an avenue to explore philosophical ideas.

Through this perspective, I do not constrain myself by trying to meet the demands of practicality. I don't even aspire to create a formal model. In fact, I shy away from that. Instead, I picture my method of exploration as nurturing openness, understanding, and flexible ways of approaching the world.

2. Don't Be Limited by Commonsense Beliefs

This concept can be a little challenging to understand. With it, I encourage thinking about human beings and hypnosis without using the usual concrete models that we rely on to understand most things.

By doing this, I'm pointing to something elusive but nevertheless tantalizing and meaningful. It can't be expressed with words.

To begin, let's examine a few of the verses in poet Walt Whitman's iconic book *Leaves of Grass*.[2] In it, he pushes us to think about the world in a way that enhances our understanding of ourselves within the world rather than one leaning toward utilitarian common sense:

> I have heard what the talkers were talking, the talk of the
> beginning and the end,
> But I do not talk of the beginning or the end.
> There was never any more inception than there is now,
> Nor any more youth or age than there is now,
> And will never be any more perfection than there is now,
> Nor any more heaven or hell than there is now.

In these lines, Whitman is grounded in a nonlinear view of life. He throws out typical notions of things beginning and ending. This is, of course, a radical departure from most commonsense perspectives. To him, life might be seen as a continuous whole, as he describes youth and old age occurring simultaneously.

Further, as in some Eastern approaches, he nudges us toward experiencing in terms of the present moment and not being limited by mistaken commonsense beliefs about the world.

By taking inspiration from Walt Whitman, many start to open themselves up to a broader view of their minds. Seeing ourselves, and the world, from such a nontraditional lens can thus help us reimagine ourselves not as within, but as integral parts of, the world.

My departure here from concrete, linear, and reductionistic models stems from my concern with a widespread tendency to explore the world only as if it were composed of separate components, sometimes interacting yet fundamentally isolated. This narrow perspective not only influences both commonsense ideas about hypnosis but more broadly exemplifies many myopic, commonsense ideas about our position within the world.

3. Embrace the Intangible, Accept Discomfort, and Relish Ambiguity

In another of his works, Whitman takes us a step further. He reminds us that words can only lead us in a particular direction. They do not necessarily spell out a clear destination or offer an entirely accurate description of anything.

This poem (again found in *Leaves of Grass*) is called "Shut Not Your Doors." A snippet reads:

The words of my book nothing, the drift of it every thing,
A book separate, not link'd with the rest nor felt by the intellect.

Whitman emphasizes that his focus is not on the specific *meanings* of his words but on where they point the reader. To him, intellectual analysis, focusing on the exact meaning of the words, only goes so far.

Whitman's writings echo those of some early Eastern philosophers. Consider the Buddhist Śūraṅgama Sūtra, which teaches:

Suppose someone is pointing to the moon to show it to another person. That other person, guided by the pointing finger, should now look at the moon. But if he looks instead at the finger, taking it to be the moon, not only does he fail to see the moon, but he is mistaken, too, about the finger. He has confused the finger, with which someone is pointing to the moon, with the moon, which is being pointed to.[3]

The lesson, once more, is that language is only a way to find, or perhaps point toward, the truth, but we should not get too attached to, or distracted by, words. Both Whitman and the ancient Śūraṅgama Sūtra therefore take us even further away from concrete, model-making thinking. They lean toward new ways of thought and rely on words that do not necessarily have to capture anything tangible to bring insight.

Like the uncertainty Walt Whitman's poem creates, the mysteries surrounding hypnosis, mind, self, and consciousness push us to question our beliefs and adopt new perspectives. That's what Anna did when she agreed to see herself as one person rather than two. Such mindsets force us to embrace ambiguity and perhaps even point toward new avenues for self-discovery.

If you are uncomfortable taking this direction, that's good. Nourish the discomfort. Learn to love that feeling, knowing you will never get bored if you don't think you already have all the answers. For me, this attitude has been a step toward thinking more freely, that is, without relying so heavily on concrete models or preconceived destinations. And, like Whitman, by only pointing in a direction rather than describing it, I'm hoping to guide us toward liberation from more common, utilitarian, and comfortable ways of thought.

As we proceed, remember that ambiguity and uncertainty can be wonderfully enticing and helpful. For example, remember the last cliffhanger you saw in a video. Not knowing what was going to happen next might have guaranteed that you immediately went to the next episode. Further, if you better see the limitations of your beliefs about yourself, you perhaps will develop an interest in deeper explorations into who you really are, and even look for the new directions this perspective could point.

4. Let Models Fade Away

Explaining something in a book, or by any means, entails anchoring it within a model. We often express ideas using words, which themselves are little models of reality. However, I'm now challenging this usual method of explanation in hopes that it is possible to transcend these ingrained and constraining linguistic frameworks.

Naturally, there is an inherent paradox in trying to use words to describe what I refer to as "no-model thinking." This is because I, like Walt Whitman, can only convey the "drift" of "no-model" with other models—that is, with words. It's like trying to explain silence by playing a song. It's difficult to imagine anything more challenging than using words to describe a form of thought that transcends the limitations of language.

And so, my use of words is a misleading necessity. Emulating Whitman's approach, my own words are used not so much to describe specific concrete processes or things but to point to something beyond the confines of language.

When reading my explanations, remember Walt Whitman's explanation that the drift of his poetry means everything and his words nothing. He does not fixate on particular words. Instead, he turns toward the overall emotional significance of his poetry. By changing your focus like this, you too might discover a new type of realization, one less constrained by past belief.*

BRINGING IT ALL TOGETHER

Not Thinking of Hypnosis, or Ourselves, as Being Limited to Our Heads

By now it should be clear that seeing hypnosis (as well as ourselves!) as centered in our heads, while useful, is only a limited view of reality. We can also think of hypnosis more abstractly and in the interconnected sense of being *woven into* our cultural norms, customs, physical environments, and, literally, every aspect of our existence.

From this perspective, hypnosis is not *caused by* relationships between people and objects, as when, for instance, a hypnotist "hypnotizes" a client.

---

* A similar helpful dynamic, that is, moving away from the precise meanings of words, can occur in hypnotherapy. This can happen when therapy moves away from focusing on specific word meanings and toward the more abstract significance of the thoughts and memories (true or not) that can come forth.

Rather, it is *part of* these relationships. It is, among other things, an integral part of society and culture, not their by-product or result.

Here in the West, we habitually struggle to appreciate this more expansive and interconnected view, partly because we are so used to imagining a self in our heads directing our actions. Our Western models of hypnosis sometimes, but not always, focus on this ostensible self, usually as the producer of something—consciousness, ideas, beliefs, intentions, and hypnotic trances, among much else—often with the assistance of a therapist, professional norms, a culture, and a society.

I propose an alternate perspective. It is one that pictures ourselves, our societies, and our cultures not as creators of hypnosis but *as inseparable components of hypnotic experiences.*

Notice how this mirrors the view that colors in a rainbow do not exist only in the sky; they appear in the myriad overlapping interactions of air, water, light, eyes, brains, position (someone needs to be in the right location to see a nearby rainbow), and a variety of other influences. All of these are required for the experience of colors to arise.

With this thinking, hypnosis, like a rainbow,* is not a separate thing confined to one specific location, such as someone's brain. It occurs in many places at once—or, more accurately, in no one place at all. We are thus obliged to observe that hypnosis *is* partly in people's heads after all, since it arises in a multiplicity of interactive locations—but it is not *only* in people's heads.

In line with this thinking, hypnosis can be viewed as a relationship between a therapist and a patient, *and* as a relationship between an individual and their culture, *and* as a relationship between the "inside" mental world and the "outside" world, *and* (as some Eastern philosophers recognize) as interactions with and among everything that exists.

Think back to when Antonio asked Anna to imagine either a fly landing on her cheek or a car inside his office. He summed up her responses using his "hypnotizability scale." This brain-centered model of hypnosis measured Anna's responses—as well as what Antonio thought happened *in her head* to create those responses. It reinforced the conventional (often productive) brain-focused model that the inside of a human head is where hypnosis occurs.

---

* Some readers will wonder: "Where do bad rainbows go?" The answer is prism. But it's a light sentence, and it allows time for reflection.

However, that model is not designed to gauge the hypnotist's, or Anna's, or their culture's proclivity to define certain behavior as signaling hypnosis. In fact, it might be just as useful, in a narrow sort of way, to predict the degree of trance by giving the *hypnotist* and the *society* tests, each made to measure how likely they are to define certain behaviors as signaling hypnosis.

Simply put, "hypnotizability" tests also show how well a patient's behavior matches a society's current definition of what hypnosis is. And so, these tests tell us at least as much about the people who design them, as well as their cultures, as they do about the people being tested.

Antonio helped to spread his own concept of hypnosis to Anna, partly with the scale he used to define it. Society also nudged him in that direction by defining certain behaviors in a particular manner—as evidence of hypnosis—rather than, for example, calling them imagination, dreams, spiritual guidance, unconscious forces coming to light, or evidence of willpower.

After embracing the idea that hypnosis goes beyond our heads, we once more see ourselves more broadly. Hypnosis, in fact, quickly becomes a tapestry of relationships, each transcending the commonsense limits of our bodies. Once we adopt this all-encompassing view, we have turned away from brain-focused thinking. Instead, we have welcomed a broad vision of our personal links with all that exists. By doing this, we not only become awestruck by the connections that make up our lives but also gain a far deeper sense of unity with the world.

Remembering That Utilitarian Thinking about Hypnosis, and Ourselves, Only Goes So Far

This concept invites us to explore ourselves beyond the confines of utilitarian thinking. It encourages us to see all behavior, including that which we label "hypnotic," in ways other than through a lens of practicality.

Think, for instance, about how we create the world. We generally think about it based upon practical utility and the subjective meanings upon which we widely agree.

For example, we productively picture some animals as "pets" rather than using concepts that describe a property inherent in the animal, such as warm-blooded. Instead, the category we use describes a relationship between an animal and its owner(s). Using the category of "pets" thus helps control our interactions with certain animals.

Remember the insights of the philosopher Sheng-yen. He observed that sensations (such as colors) perish when they are separated from lighting conditions, someone to observe and interpret the colors, and our own biology. Similarly, a broad view of hypnosis is that it occurs in no one place but in multiple locations as many causes come together. Seen in this way, hypnosis does not exist as an isolated entity—*and cannot be fully understood by either dissecting it or confining it within a boundary.*

This holistic perspective, unfettered by reductionistic thinking, greatly reduces any need to guess about what happens during hypnosis, unseen, in people's heads. There is, for instance, less need to wonder—or to try to discover—whether people incorporate others into their own sense of self.*

This perspective can also lead us to a broader, more abstract understanding of the world. We can recognize that anything we observe only acquires its qualities in relation to many other "somethings." For example, there cannot be the top of a tree without a bottom to the tree, or the soil, or water, or air, or sunlight, and so on. Similarly, there cannot be hypnotized people unless we have a social consensus about what hypnosis is, how nonhypnotized people behave, under what conditions hypnosis can and cannot occur, etc.

To better comprehend this reasoning, first remember the old joke:

Q: Where did you lose your shoes?

A: In the basement.

Q: Then why are you looking for them in your backyard?

A: Because that's where the light is.

The illogic of this humor applies to some Western models of hypnosis. That is, if our goal is to achieve a better understanding of ourselves, we've sometimes been looking for hypnosis in the wrong place. As we know, many have focused their search inside the head, which is where many of our scientific Western models of the world point and where we are often most comfortable looking. The narrowest of such views picture hypnosis as something we can put under a metaphorical microscope. These are the models in which our ideas of the world as composed of separate, perhaps interacting, parts lead us to seeing hypnosis mostly in isolation.

---

* Recall that we would do this if we speculated about Anna incorporating Antonio's hypnotic suggestions into her own sense of "self."

Concepts such as self, mind, hypnosis, and consciousness are sometimes models of our insides that seem to fit together into a machinelike system. But they are not actual entities that can be located or pinned down. We imagine (or infer) that they exist, just like the insides of a watch we cannot open—or, perhaps more accurately, inside a watch that we can open but whose envisioned key parts we cannot find.

This is not to say that there is not much benefit in exploring what happens inside people's skulls. Research into brain activity has given us many useful insights into brain function and psychotherapy. But let us not conclude that such research can ever give a full picture of hypnosis, perception, consciousness, or what it means to be human.

To try to conceptually grasp the enormity of such experiences as love, spiritual insight, hypnosis—or, for that matter, the nature of mind, self, or what it is to be human—is like trying to pinpoint a single cause or source of a Mozart concerto. Do we throw a conceptual lasso around Mozart's talent, his training, his inspiration, the invention of the piano, his patron, a good teacher, chance, a tasty meal on a sunny day, the listener's appreciation of his music, or something else entirely? Any accurate picture is endlessly complex and an amalgam of these factors, as well as many more.

But this is hardly cause for despair. Indeed, this awareness is profoundly helpful. It helps us to avoid falling into rabbit holes or pursuing imaginary answers at the end of conceptual rainbows. And it encourages us to consider all models, scientific or otherwise, not as ultimate explanations but as potential guides to help us accomplish whatever task we are presented with.

This approach to hypnosis—and, perhaps, Eastern ideas in general—might not sit well, especially among people looking for scientifically useful answers. When we include literally everything as part of hypnosis, this may at first blush appear to cast the concept in hopelessly complex terms. Perhaps that's true. In fact, however, problems can also come from the *opposite* direction: keeping things concrete, simple, and with clear boundaries. *These* are the perspectives that are ultimately obstacles to better understanding—or perhaps a greater or truer vision—of what we might call the human mind.

Opening Doors and Shifting Our Perspective

Obviously, shifting perspectives using concepts like those above can change us. For example, once we stop imagining hypnosis as something that goes on only inside our brains, we can start to let go of other narrow thinking and myopic views about ourselves and the world. Perhaps we can be more aware, and more curious, about our brains' power to mislead us through their interpretive processes. And perhaps we can see beyond our commonsense view of the world as an enormous collection of things (including people), each existing in relative isolation, with clear-cut boundaries.

Freud also shifted not only his own perspective but those of much of society. Even when wrong, his innovative ways of thinking about the world made an indelible mark on society. We still use some parts of his approach in therapy. For example, we now often explore past experiences to explain and address current challenges. His new concepts have also helped us both to create and interpret art and to identify their deeper representations of the human condition. And, of course, he has shaped many views of parenting.

You'll remember from chapter 3 how malleable people's models of *themselves* were. In the study involving a mannequin's arm, subjects routinely experienced that false arm as their own and incorporated it into their own sense of self. The boundaries between themselves and the world outside were shown to be fluid.

Such shifts, each broadening our perspectives, are far more than exercises in thought. They can also foster the essential attributes of growth, well-being, curiosity, and openness. These characteristics are crucial in adapting to our ever-changing world.

Awestruck

Broadened "no-model" perspectives, such as those that view humans as inherently connected with the world around them, are especially likely to create a sense of awe. Psychologists now think that awe can sometimes be a transcendent and incredibly powerful experience, even a life-changing event, triggered by moving our attention away from ourselves as separate entities and instead seeing humans as part of something both vast and extraordinary.[4]

Not surprisingly, these awe-inducing events can have many benefits. These include enhanced feelings of connectedness with the world, greater charitability, increased skepticism, improved mood, and less desire for material possessions.[5]

Hypnosis as a Reflection of Society

Using this different lens also encourages us to look more closely at the societies in which various ideas about hypnosis emerge. How a society defines any experiences or behaviors, such as those seen in hypnosis, reveals its morals, customs, and beliefs.

For example, societies in which science is prominent naturally tend to develop mechanistic psychological or neurological models of hypnosis. They also often view hypnosis as a medical technique to access hidden conflicts in an unconscious mind, or as a reflection of neurological anomalies brought about by unique social demands.

This medical perspective legitimizes hypnosis as a therapeutic concept and method rather than as a tool for stage performers and charlatans. It gives people access to hypnosis as a healing strategy. It stimulates productive evidence-based scientific research based on that method.

At the same time, however, such an approach can stigmatize people by defining them as patients and their behavior as illness—rather than seeing their behavior as a common, and often helpful, way of changing perspectives. This can lead to unnecessary, and potentially harmful, medical treatments, as well as to people's unwarranted concerns about their own well-being or sanity.

In other cultures, the very same "hypnotic" behaviors are considered signs of faith, revelations, miraculous healings, contact with another realm, or even the power of God. Some, for example, think that "visions" show special abilities connected to a spiritual realm.

In short, science, sociology, psychology, and philosophical thought deeply intertwine. In doing so, they reveal a far richer and more profound picture of humanity than simply "The therapist hypnotized the patient, and the hypnosis cured her."

Seeing this complexity helps us understand and appreciate the diversity of human experience. It highlights the beliefs that both drive advancements and trigger problems around the world. In the chapters that follow, we will explore this richness further.

CHAPTER FOURTEEN

THE MYTH OF A HEALTHY AND SINGLE "I"

We have learned that when people feel disconnected from their own thoughts, emotions, and sensations, mental health professionals often refer to this experience as dissociation. These professionals typically describe dissociation as a sense of feeling distanced from, or losing control over, one's personality, emotions, thoughts, body, or speech.

On occasions, dissociation can be linked to traumatic events. After such occurrences, some people describe sensing an internal division, at times feeling as if they have multiple personas, which can lead to a sense of internal fragmentation. In psychology this symptom can be referred to as a problem with self-understanding.

People who dissociate sometimes report they can feel the periodic emergence of an alternate self that seems to "take control" away from a primary self. Interestingly, this back-and-forth phenomenon resembles patterns observed during some physical interactions. These include couples taking turns leading on a dance floor, or drivers switching seats in a car.

Today, the prevailing belief is that those who claim to have more severe forms of dissociation, including multiple personalities, are considered mentally ill—i.e., that they have a medical condition. Indeed, in the DSM-5,[1] the go-to manual for diagnosing mental illnesses, "dissociative disorders" are treated as medical issues, just like broken bones. This medical model is useful. It helps spot problems and guides treatment for people in distress.

This medical perspective also commonly presumes that a normal, healthy human being should always operate *as if* there is a single, definable personality in control. However, this kind of view of what it is to be human is also limiting and incomplete. We should therefore ask if there are other broader, more inclusive concepts that occasionally might serve us better.

Avatars and Egos

But what about changes in our sense of self that are brought about by digital worlds? These transformations in how we view ourselves might happen online, such as when we use gaming avatars, social media profiles, and anonymous forums. In settings such as these, we often show traits that are not characteristic of our typical behavior. We might seem to temporarily become, both to ourselves and others, almost different, or even multiple, people.

Cultural perspectives also impact how we perceive different identities. Some philosophies, for instance, interpret a detached or a sense of a second self (regardless of the cause) as a sign of heightened consciousness, or even "transcendence." Clearly, our views regarding multiple identities or "selves" have strong roots in culture.

The default model, where we behave as if there is one person inside our skin, is what I call an *associative* model of personality. This model posits that everything we perceive and do is filtered through the lens of this single internal persona.

Here is our conundrum: According to some thinking, a fully functional and healthy human being acts as if there is only one "I" within them. *Yet, as we know, such an "I" remains elusive, can never definitively be found, and can even be thought of as a fabrication.*

Furthermore, depending on changing circumstances, it can be beneficial to act *as if* we were more than just one individual. And, as we have seen with selving, at any given moment our sense of self is malleable, constantly shifting based on current circumstances and relationships. We therefore know that firm boundaries for this self, which might define the difference between "I" and "not-I,"[2] are elusive.

To clarify, I am not suggesting that the default "associative" model of just one personality should be discarded. Not at all. I'm simply again emphasizing how *the ways in which we think about ourselves are malleable and adaptive*—and that these are hugely positive human attributes.

Now, let's revisit some key points from our previous discussions.

First, our brains are enormously flexible and adaptable—and able to entertain diverse models.

Second, we have no concrete evidence supporting the existence of a single enduring "I" within us.

Third, we have very little compelling evidence that there are multiple "I"s inside each of us, either.

Fourth, social and cultural expectations and norms—as well as the self-deceptive interpretive process of our brains—can lead us to unfounded conclusions.

With these ideas in mind, we can see that dissociation, such as experiencing a sense of having different personalities, is not necessarily a *pathological* condition. In certain contexts it causes problems—or is, at least, a signal of psychological distress. But at other times, dissociation shows our capacity to navigate life's difficulties.

The Chameleon Within

I sometimes think about what happens when, during psychotherapy, we scrap the old-school notion of a permanent, internal "self" and turn to the concept of a fluid "self." For therapists who have already embraced this perspective, Anna's story isn't a shocker. Instead, they probably observe that it's like an amplified version of what we all do every day.

Here's a fun way to look at it. We're all chameleons, almost seamlessly accepting different roles depending on the setting and the people we meet. You might be a no-nonsense executive leading high-stakes negotiations during the workday, yet quickly transform into the life of the party when hanging out with friends.

In a similar vein, Anna created Sarah to deal with her struggles. She did this even though Sarah was built on false memories. Seeing this, psychologists who embrace a fluid concept of the "self" understandably can think of Anna's experience as more of a form of self-adaptation rather than a gigantic deviation from the norm. Anna's hypnotherapy adventures aren't shrugged off as a weird exception but accepted as one (likely misguided) strategy for personal growth.

In this view, healing becomes more about learning to accept the shifting aspects of ourselves than about nailing down a relatively static self. This attitude might help other people with concerns like Anna's. This is because they would no longer see themselves as being people with unique psychological conditions.

In fact, much of the angst people feel from a sense of shifting selves can be attributed to many societies' stiff and unyielding views of identity. This rigidity can make some people have an even greater sense of shame. Sadly, it can also trigger more mental health issues. In this manner, societal views can be seen as self-fulfilling prophecies: Society causes people to have distress, then refers to this distress as evidence to justify why it stigmatizes the problem. But, once again, when we adopt a view of a healthy changing self, such concerns can vanish.

Dividing People into Multiple Selves: When Help Can Harm

But let's not get overly enamored with the idea of accepting multiple selves without addressing some concerns. The type of treatment that was given to Anna is sometimes called *ego state therapy*.[3] This therapy is all about helping people examine and perhaps untangle the various "parts" of their identities. Think about an "ego state" as a consistent way of thinking or acting that appears as a distinct aspect of a person's personality. Many even liken these distinct aspects to temporary additional personalities. As we saw with Anna, the therapist helped create the temporary entity "Sarah" and then performed a type of counseling with both, almost as if the alternate "ego state" in Anna was a separate person.

Let's talk about the dangers of readily accepting, or encouraging, different personalities in one individual. It is tempting to label any changes in thinking as unique and separate ego states. Here, if a therapist noticed a patient behaving differently from what that therapist had come to regard as the client's "norm," that behavior could be thought of as a separate ego state, or personality. This can create confusion. Imagining that different strong emotions signify different ego states, or even indicate separate "selves," sets you up for a wild ride.

This approach also suggests that only one personality cannot cope with life's ups and downs. And the belief that patients could benefit from "compartmentalizing" different parts of themselves might imply that a unified "self" is faulty. With Anna, for instance, what if she concluded that her core personality wasn't enough? Some patients (perhaps like Anna) even find it belittling that the therapist must drag out different personalities/ego states to help them cope.

There are some patients who come to therapy who already have a shaky sense of self. Fragmenting them into different ego states can make their lives even rockier. Instead of helping them integrate different "parts" of themselves, therapy could pull them into different directions.

To demonstrate the dangers discussed above, let's think about Frank, a teenager who began ego state therapy already having a shaky sense of his own identity. He had extreme emotional swings. Sometimes he had intense self-doubt, and sometimes his self-esteem was inflated. His therapist, Dr. Singh, regarded these swings as though they were different aspects of Frank. Sometimes there was "Confident Frank," and sometimes "Insecure Frank" made an appearance. At first, Frank felt validated. However, after some time, these labels turned into barriers. They made his emotions seem even more fragmented.

By thinking about these "ego states" as if they were semi-independent people, Frank started to lose confidence in what he previously thought of as his "core" self. He believed that, after therapy, he would not be able to handle life's stressors without thinking about himself in terms of "smaller" pieces. He worried, at least until he received reassurance about his abilities, that later he might not be able to guide his life without convening an internal "council" of ego states.

Frank's experience is a cautionary story. This ego state approach, while maybe helpful for some people, should be used rarely, if at all, with full knowledge of the downside. Most ups and downs don't warrant the label of an ego state. Instead, helping to nourish patient feelings of having a sense of unity is usually a more helpful goal.

And so, whenever we challenge traditional notions of self, we should use caution. There's no one magic fix for everyone, after all.

CHAPTER FIFTEEN

THE MYTHS OF CONSCIOUS CONTROL AND SELF-UNDERSTANDING

One commonsense model of humans is that our actions occur consciously and intentionally in response to the thoughts, impulses, and sensations that fill our awareness. We think we peer inside ourselves, observe our own urges, sensations, and thoughts, and use these observations to explain our actions. This belief—that our deeds are driven by the conscious nature of our experiences—is rarely challenged.

For example, we think we buy shoes because we're drawn to their look and fit. We believe we locate a pencil in a jumble of objects by first picturing it in our mind. We see a mushroom in front of us on a woodland path, then think that we decide to avoid it to preserve nature. But these conclusions, as we will see, might not be true.

Anna had similar beliefs, at least at first, about the causes of her actions. At first, she thought that her overeating stemmed from the newly unearthed conflicts Sarah had revealed. But she was soon surprised when she learned that these "reasons" for her actions were only recent, fleeting ghosts. She eventually saw that these explanations were unlikely culprits for her long-standing issues with her weight. This profound insight challenged her earlier beliefs about her eating habits.

Most of us can recall times in which we acted out of character, did things that "weren't like us" at all—or did something that we simply couldn't (and perhaps still can't) explain.

In each of these examples, involving physically healthy people, beliefs about what motivated our actions were later either debunked or missing.

Seeing without Awareness

The neurological problem of cortical blindness[1] throws another wrench into the commonsense model that we do things because of the nature of what we experience. People with this condition have functional eyes, but the part of their brain that generates awareness of their surroundings is damaged. This brain tissue normally sends visual signals to other portions of the brain that then make our visual experiences, like recognizing a pencil or a pair of shoes.

But when the tissue that relays these signals is impaired, the signals from the eyes don't reach the image-creating parts of the brain. In short, somebody with cortical blindness can have eyes in tip-top shape, but the person does not "see" visual images in the way we'd expect.

It is therefore no surprise that if you show someone with cortical blindness an object—let's say a large toy—and ask them where it is on the table in front of them, they might say that they can't see either the toy or the table. And, so far as they can tell, they don't.

Yet if you then ask the person to guess its location, they tend to answer correctly. *They often appear to "see" it without knowing that they "see" it.* This phenomenon—which has been repeatedly confirmed in experiments—is sometimes called *blindsight.*

In these instances, people's correct answers about the location of the toy conflict with the commonsense model that the "phenomenological" properties of seeing the toy allowed them to give its correct location. They give the right answers without being aware of any image to inform their responses.

When researchers ask these people how they knew the toy's location, most say something like, "I don't know; I could just sense it somehow," or "You asked me to guess, so I did."

With this phenomenon, then, the seemingly obvious truth—that we do things only because of our conscious realization of what is around us—is not correct. Someone with cortical blindness simply cannot "look inside" to learn why they were able to pinpoint the toy's new location.

---

* Philosophers (and some neuroscientists and mental health professionals) use this "blindsight" phenomenon to distinguish between two types of consciousness. The first is called *access consciousness* and pertains to information that can guide our actions, such as reasoning or speech. The second is *phenomenal consciousness*, which refers to our experience of what things "are like," such as how they appear. Adherents of this two-kinds-of-consciousness model say that visual information, such as the color of an object, is in some sense first available to the brain for processing via access consciousness, but later on might not be accessible to phenomenal consciousness, i.e., what we think we see. Although our experience of the image "yellow ball" is not the cause of the correct response, these folks can argue that the underlying information (that we don't experience) about a yellow ball *is* the cause. They think that these two types of consciousness exist because the brain does process certain information, but not at a conscious level.

Reevaluating Human Exceptionalism

Instinct blindness reveals a limitation in examining our own *thoughts* to discover why we act. In 1890, about three decades after Darwin published his *On the Origin of Species*, psychologist William James argued in *The Principles of Psychology*[2] that humans, much like other animals, are guided by instinctive behavior. Indeed, James believed that humans' adaptability comes not so much from rationality but from the complexity of our instincts.

Other thinkers and researchers believe that we humans are often blind to our instincts precisely because those instincts operate so well. The theory goes that all animals, including humans, don't need to know the causes of their actions. They just need to act effectively. From this perspective, we all engage in a variety of actions fine-tuned via natural selection over countless generations.

For example, we admire a goose's navigational skills during its winter migratory flights, or its flock's ability to fly in an aerodynamic wedge formation. Yet geese aren't crunching numbers or following a set of logical instructions as they fly. They don't need to. Furthermore, it's doubtful that a goose can, or needs to, reflect on its own navigational reasoning. The journey's success does not require this level of insight.*

Similarly, a human piloting a hang glider, or a center fielder catching a high fly ball—or a dog snagging a Frisbee in its mouth—doesn't rely on mathematical calculations or "higher intelligence" to succeed. We instinctively jerk our foot off a sharp rock, seemingly without thought. It is only afterward that we might (possibly mistakenly) deduce that we "decided" to withdraw our foot because of the experience of pain.

Like other animals, then, we humans often can't pinpoint how or why we do something. Indeed, perhaps we don't even know how or why we are *able* to do it.

Moreover, we don't *need* to know such things. *Nature favors survival and reproduction, not necessarily accurate knowledge about how or why we do anything.*

In fact, this unawareness can be a crucial survival skill.

---

* Descartes noted that if birds migrated because of intelligent, rational behavior, then they would have to be very intelligent creatures indeed. If that were the case, however, we might then expect to see them apply this intelligence across a wide range of abilities. But, of course, they do not.

Think about the last time you used a computer to write an email. You needed only to know what was on the screen and how to operate the mouse and keyboard. Yet in composing that email, your brain accurately executed countless swift and detailed calculations, each without your awareness. But what if you *were* aware? Your brain would be so bombarded with information—information that you didn't really need—that you would perhaps be too overwhelmed to write that email. In this sense, ignorance is truly not only bliss but necessary for our daily functioning.

Instinct blindness does more than obscure the reasons that we do certain things. We like to think that we're evolution's crown jewel, superior to all other species. However, recognizing that many of our actions are governed by automatic processes, rather than by our cleverness, is humbling. Maybe that's not such a bad thing.

Looking beyond Our Choices

In psychology, there has been a long and ongoing debate about how much our thoughts affect our behavior.

B. F. Skinner, for example, emphasized that our underlying neurological processes, rather than the nature of our "phenomenological" *experiences* of thoughts, were the primary drivers behind our actions. According to Skinner, our observable behavior, such as grabbing a drink of water, might come more from our nervous system's interaction with the environment than from our thoughts.

For Skinner, a thought is *related to* the neurological system but is not necessarily the sole (or even a contributory) *cause* of any action that follows. He explained it this way: "We may think before we act in the sense that we may behave covertly before we behave overtly, but our action is not an 'expression' of the covert response or the consequence of it."[3] As we saw earlier, he viewed thoughts as preparatory exercises and internal rehearsals rather than as vital precursors of actions.

Several models of the brain are at odds with Skinner's thinking. Perhaps the most prominent of these include *computational models*. These liken the human brain to "a system of organs of computation designed by natural selection to solve the problems faced by our evolutionary ancestors."[4] In this view, the human brain is essentially a computer that evolved to aid our and our ancestors' survival.

Researchers in favor of computational models sometimes think about the mind as being divided into modules, each shaped through (or at least influenced by) natural selection. One module turns visual information into useful perceptions. Another helps us solve math problems. A third helps us predict others' actions. Some also believe that parts of the brain evolved highly specialized modules, like one focused on catching people who cheat.

Advocates of computational models view knowledge, beliefs, and desires as physical states of the brain that prompt specific behaviors. Hence, if we adopt a computational standpoint, understanding someone's knowledge, beliefs, and desires becomes deeply important. This is of course much different from Skinner's approach. Steven Pinker[5] eloquently explains with this simple scenario:

> Why did Bill get on the bus? Because he wanted to visit his grandmother and knew the bus would take him there. . . . Beliefs and desires are the explanatory tools of our own intuitive psychology, and intuitive psychology is still the most useful and complete science of behavior there is. To predict a vast majority of human acts—going to the refrigerator, getting on the bus, reaching into one's wallet—you do not need to crank through a mathematical model, run a computer simulation of a neural network, or hire a professional psychologist; you can just ask your grandmother.

In essence, computational models posit that each of us is substantially aware of our knowledge, beliefs, values, and thoughts; that our brains can examine these (acting as both the object of examination and the examiner); and that these mental events are what cause our actions.

By now, however, we can wonder if these models are always the best way to picture ourselves.

Imagine that your friend knocks his coffeepot off the counter. You can ask him why that happened. But you would be wise to first remind yourself that our culture has probably heavily influenced him—and that his brain, like yours and mine, has a long history of making up explanations. (Remember the left-brain interpreter?) You might be wise enough to trust this answer the most: "I don't know. Should I make a new pot?"

Cultural beliefs, of course, evolve and vary across groups. One age blames demons for mental illness, and another implicates body humors. In our own, we often pin the blame on neurological problems.

Different people will have varied ideas about why they do things, depending on their personal experiences. One friend finds comfort in prayer before a dental appointment, believing that divine intervention calms their nerves as the dentist works on her mouth. Other friends credit their relaxation to imagination, willpower, or faith in the dentist—or even a trance induced by a therapist the previous week. Each of these friends might—from their own experience and perspective—be telling the truth. But we should be cautious before accepting any single answer as complete and accurate.

Adaptive Dishonesty

Here's another wrinkle: As we have seen, our ancestors evolved in communities where disseminating—or pretending to believe—disinformation sometimes proved useful. This still holds true today. Whether it's a seasoned realtor feigning excitement about a home's "curb appeal" or a student who is caught cheating feigning remorse to avoid expulsion, we often witness disinformation in action. Can we really take these people's responses at face value?

Or consider the commonplace act of falsely implying that you have respect for your colleagues when you know that some of them are frauds and incompetent. Recall here Steven Pinker's insightful explanation that "our brains were shaped for fitness, not for truth."[6]

Pinker also discusses "sham emotions"—the display of emotions not truly felt at that moment (p. 405). According to Pinker, this behavior evolved because it is often successful in influencing others.

Of course, this doesn't mean that we can't glean insights from talking with friends. But we shouldn't treat their—or anyone's—answers as infallible windows into the workings of their brain.

In brief, we can't base science on people's common sense and fabrications.

Intentions and Illusions

It makes intuitive sense that first we decide to do something, and then we push a mental "execute" button and do it. Certainly, this is how we *experience* many of the decisions we make and the actions we perform.

We experience things this way because—so far as it seems—we look inside our own brains, observe our motivations, watch as we will an action to occur, and observe the action as it unfolds.

In many (if not all) cases, however, this is not what happens at all.

Common sense–shattering research by Benjamin Libet[7] has revealed that we often begin to act *before* deciding—or becoming aware of deciding—to do so. In a series of experiments, Libet attached electrodes to people's heads to measure electrical activity in multiple parts of their brains. He then asked subjects to press a button whenever they felt the urge to do so.

Libet looked at the order in which different parts of people's brains fired, both before and after they pressed the button. He discovered something counterintuitive: People did not experience themselves deciding to press the button until almost half a second *after* their brain had issued the signal to press it. According to his findings, *their action preceded their decision to act.*

While Libet's methodology has been challenged, some of his general findings were supported years later by Chun Siong Soon[8] and others. They used more advanced technology to further study when the brain started "preparing" to act and when people reported becoming aware of their intention to act.

In Soon's study, subjects were asked to decide to press one of two buttons and then to do so immediately after they had made their choice. The researchers then examined the activity in different brain regions that would predict the button presses before they occurred. They learned that they could use this brain information to predict with great (but not perfect) accuracy which button the subjects would later decide to push. More strikingly, they found that they could predict which button subjects would press as many as *10 seconds* before the subjects themselves were aware of having made the decision.

We believe that we can observe our thoughts—i.e., read our own minds—and thereby discover why we do things. Yet Libet and Soon have shown that we can act *before we even know that we have begun to act.*

Therefore, our brain shows itself once more to be a brilliant storyteller. It weaves explanations to make sense of things after the fact. We have seen this storytelling before, of course, with both rationalizations and confabulation.

In these experiments, even though people's brain activity regularly preceded their decisions to act, they did not *perceive* the sequence that way. They told the experimenters—and seemed to honestly believe—that first they decided to push the buttons and then they pushed them.

Libet's colleague Walter Freeman* describes this reality beautifully:

> The private experience of self, the "ego," is invariably half a second behind, always justifying, explaining, rationalizing, and claiming credit by virtue of the capacity to back date. . . . The illusion is not of the existence of the ego, but of the ego as being in control of the self.[9]

We see again that our commonsense models of human self-understanding can be wrong. They do, however, suggest some more deeply pertinent questions. We'll look at some of them in the chapters that follow.

---

* Libet and his colleagues have edited a book on this subject called *The Volitional Brain: Towards a Neuroscience of Free Will.*

CHAPTER SIXTEEN

AUTONOMOUS SELF-CONTROL

Our human sense of free will has been debated over centuries. Can our brains direct our actions independently of their surroundings? This all comes down to the basic question of whether we have free will—and it's an important question. But how can we possibly answer it? How might we create a set of conditions and circumstances in which a human subject's decisions are independent of their surroundings?

We would be wiser to begin by looking at what creates our impression that we have free will.

The Ouija Board Phenomenon: Interpreting Free Will

In 1999, Daniel M. Wegner and Thalia Wheatley[1] set up a clever experiment using one experimental subject and one person who only *pretended* to be a subject, but who was really a secret accomplice of the researchers. Working together, both the subject and the secret accomplice simultaneously pushed the same computer mouse across a table, as if it were the planchet on a Ouija board. As they moved the mouse, the cursor on the computer screen moved over and around various images of objects.

Every so often the secret accomplice would halt the mouse on the table. Sometimes just before the accomplice stopped the mouse, the subjects heard the experimenter present a word that corresponded to an object at the cursor's current location on the computer screen as the mouse came to rest. For instance, just before the accomplice stopped the mouse by an object on the screen, the subject and accomplice might have heard the name of that object. The actual subjects, however, were told that the word was intended purely as a minor distraction and that it should not affect their mouse movement.

The point of the experiment was to determine if subjects would mistakenly conclude that *they alone* had stopped the mouse in a particular place, when in fact it had been stopped at that location by the secret accomplice.

The results were eye-opening. Findings showed that subjects often incorrectly believed that they, and not the accomplice, had stopped the mouse. This happened most frequently when subjects (1) heard the "distractor" word shortly before the mouse came to a halt, (2) heard a "distractor" word that matched the object at the location where the mouse stopped, and (3) thought that the accomplice had no reason to stop the cursor there.

In essence, these three things *together* caused the people to believe that they had freely exerted their will to stop the cursor on a named item—when, in fact, it was the accomplice who made the mouse come to a halt.

Put simply, from a scientific perspective, subjects' sense of free will *was not an accurate belief, but an interpretation.*

There are other times that we *think* that we know why we did something—but in fact we falsely assumed responsibility for something that was entirely beyond our control.

For example, late one night you go online to buy a refrigerator. You begin your search, but you're very tired, so you think: *I can put up with the noisy motor in my current model for a while. I can put off getting a new one until next week.* Then you shut down your computer and go to bed.

Nine days later, you return to the shopping website and discover that the price has gone down by almost $100. You congratulate yourself for wisely choosing to wait. In fact, however, you've forgotten that you didn't make a strategic decision to wait; you stopped your search because you were tired.

Experiments have shown that the opposite can happen as well, such as when people attribute their own movements to something else. For example, if a hypnotist tells you that your eyelids are becoming heavy, and they *do* start to feel heavy, to what do you attribute this feeling? The hypnotist's words preceded the feeling of heaviness—they were consistent with the sensation of heaviness, and there was no other apparent cause for that sensation. Therefore, you interpret the closing of your eyelids as the result of the mysterious power of hypnosis. In this model, you might feel that, during the hypnotic induction, you had no free will.

Furthermore, as we saw earlier, people do not always follow the sequence that free will would normally require. Such a sequence involves (1) making a decision, (2) initiating an action based on that decision, and (3) performing that action. Yet, remember that Libet and Soon appeared to have shown that, at least some of the time, our actions are initiated long *before* we are aware of deciding to act. If true, this would be the antithesis of free will, at least in the way it is commonly imagined.

Indeed, to scientifically prove the existence of free will, we would have to do the impossible: show that we, unlike everything else in the world, are unaffected by our environments—and, for that reason, able to operate independently of the laws of physics, biology, chemistry, and space-time. Without meeting this requirement, "will" would not be "free," because external forces would help dictate our actions.

A more reasoned scientific conclusion about how and why we do things might be this: *A brain interacts with, rather than freely directs, itself and its physical surroundings.*

Historian and philosopher Yuval Noah Harari[2] further argues that free will simply does not square with evolution. If we were indeed free to do as we wished, we would not necessarily make choices that promoted the survival and adaptation of our species. Unfettered free will would have trumped natural selection—and our species would have quickly died out.

The Limits of Autonomy

The more closely we examine free will, at least from a scientific perspective, the more it resembles outdated concepts like body humors and luminiferous ether—the theoretical substance that scientists long believed was the unseen medium through which light waves traveled. Like all of these, free will seems almost certainly to exist only in our imaginations.

As we've seen, there is utilitarian value in dividing the world. These divisions enable us to communicate and interact with one another, and to get through the day. We observe a tree and mentally separate each green leaf; then we separate each leaf into a set of attributes, such as greenness, smoothness, and perhaps even leafiness. More fundamentally, we divide the universe into two essential parts: "I" and "the outside world."

But once we discard these divisions—especially the one between "I" and "the outside world"—all confusion about free will disappears. When such distinctions fall away, there is no separate "internal" mind to exert a controlling force on a separate "external" world—and vice versa.

In the words of logician Raymond M. Smullyan,[3] "once you can see the so-called 'you' and so-called 'nature' as a continuous whole, then you can never again be bothered by such questions as whether it is you who are controlling nature or nature who is controlling you."

Let's see this even more clearly by imagining two particles so deeply connected that they stay linked, even when far apart. We previously saw that this connection means that measuring one particle's state (such as its direction of spin or its momentum) tells you the state of its companion. It's like seeing one dancer do a pirouette and knowing that the other dancer is doing the same move, no matter the distance. This is difficult to visualize or accept and hints that entangled particles might be thought of as only one object, just like we might imagine people and their environments.

Free Will and Religion

For centuries, religions have often helpfully addressed important questions about human nature. For example, religious followers have often talked about how important it is to be morally accountable for one's own actions. They also encourage us to look past a me-first mentality to instead think about how our choices impact others. These kinds of beliefs have often nurtured the development of societies that value learning, fairness, compassion, personal growth, and forgiveness.

For instance, some Eastern thinkers who follow the teachings of Buddhism believe that people have the chance to shape their own destinies through the choices they make. They also encourage people to consider the eventual wider karmic consequences of their choices. Certainly, free will is also thought of as an important part of people's eventual searches for transformation and spiritual enlightenment.

Christianity commonly embraces free will as a divine gift given to humans by God. With this, of course, comes having to bear the consequences of one's own choices. Remember that the story of Adam and Eve, for instance, showed the major costs of humans' presumed ability to freely make their own choices.

Many other religions also show profound respect for humanity in their beliefs about free will. Although their beliefs often don't dovetail with science, their positive impact on many societies, and individuals, is undeniable. It is only by having respect for both perspectives, and their interplay in each of our lives and communities, that we can gain the best understanding of the complexities in our lives.

How a Belief in Free Will Shapes Our Own Lives

Even though, at least from a scientific perspective, our notions of free will might be fabrications, they nevertheless help us survive—and even thrive. Perhaps an ability to create the *impression* that we control our own actions serves a valuable purpose. Although the idea of free will might only be another creation of our brains, paradoxically it can be an empowering and inspiring one.

Philosopher Thomas Metzinger[4] suggests that believing in free will serves various individual and community needs:

- It encourages people to compete—in a positive way—to serve a group's interests.
- It encourages people to say, "I chose to help you," allowing them to claim the moral high ground, and others to reward them for their apparent choices.
- It enables people to take credit for nearly anything they're involved in,[*] while also allowing them to blame others for their own problems, such as poverty.

Therefore, it makes sense to consider and apply the idea of free will as a social concept, one that helps us control and stabilize certain aspects of our social environment.

The *idea* of free will is thus ultimately not only compelling but sometimes even crucial to our survival and well-being. Philosophically, it is too significant to dismiss, even if it is nothing more than a social concept. All of us have a great deal of skin in it.

By now it should be evident that we need to let go of many of our common-sense models of the human brain. In many situations, we should rely more on

---

[*] Interestingly, the common model of free will has exceptions in commonsense thinking. These enable us to do the opposite: avoid blame by disclaiming any of our own free will and pointing to the presumed free will of someone else. Imagine that you catch your four- and seven-year-olds eating cookies that they took from the cookie jar against your instructions. Seven-year-old Ben points to Sarah and says, "She took them." Five-year-old Sarah says, "Only because you told me to." Whose fault is Sarah's behavior? Did she have free will in this situation? If you were their parent, would you punish her?

evidence rather than mental fabrications and be open to radically changing the ways we view ourselves and the world.

Accepting our complexity helps us better understand ourselves. It throws out, for instance, simple ideas of nature versus nurture, as if one were independent of the other. Instead, it opens doors to almost an infinite number of theories about psychology, physics, and philosophy. It even helps us incorporate spiritual perspectives into our self-knowledge.

By doing so, we start seeing the world in color, rather than in black and white. Rather than striving to pinpoint the exact causes of our actions, we sometimes might even be content to accept the intricacies of our nature.

In this book, we have already started to embark on this journey together. However, to continue, we must first challenge and dispel the single most fundamental myth about the human mind—the myth upon which many of the other myths we have so far examined rely.

CHAPTER SEVENTEEN

HUMAN CONSCIOUSNESS AND INNER PEACE

The study of consciousness has, at times, been like a 400-year-long catfight, one that arguably goes back to René Descartes and John Locke. Multiple opposing takes on it are the norm. Even today there is still no consensus on the true nature of consciousness. One contemporary writer cynically said that consciousness

> is impossible to define except in terms that are unintelligible without a grasp of what consciousness means. . . . Consciousness is a fascinating but elusive phenomenon: it is impossible to specify what it is, what it does, or why it evolved. Nothing worth reading has been written about it.[1]

Perhaps this assessment is too gloomy. Even so, in my view the enduring struggles in this area show that we might be on the wrong track.

Defining Consciousness

We might think of consciousness partly in terms of *awareness*—as the ability to freely look inside and outside ourselves and to identify an "*I*" inside and *the world* outside. By this thinking, a dog or a bird or an otter is conscious—though perhaps not to the same extent as a human—while a tree or a mushroom is not.

With this approach, I, like many others, use the term *consciousness* in a phenomenological sense, where it centers on our experiences, such as "what it's like."[2] "What it's like" can include awareness of a fact, oneself, internal mental states, having voluntarily performed an action, or even of an awareness of consciousness itself.

My main concern with such common uses of the word *consciousness* relates to a usually unspoken part of "what it's like." That issue is the question, "What it's like *for whom?*" If you speak of consciousness like this, then there is usually, at least in common* Western thought, the assumption of an "experiencer," or "I," at the core of consciousness. It is this "I" that supposedly experiences and thus is said to be conscious. This thinking aligns with the almost universal, often-Western, assumption that there is a thinker (an "I") behind our thoughts.

Thus, such phenomenological ideas about consciousness encourage a hodgepodge of *other* commonsense notions of the mind that, as we have seen, are either dubious or false:

- We possess the ability to peer inside our own head.
- We have the capacity to scrutinize our thoughts, impulses, decisions, and internal commands and to understand why we do things.
- The belief that our explanations about our experiences and the world are usually accurate, rather than after-the-fact rationalizations crafted by the brain.
- There is a disconnected "I" "observing" an outside world, which is not an integral part of all that exists.

Consciousness as a Creation

Following such conventional models, some have tried to show why a body might create something as abstract, elusive, and potentially immaterial as consciousness. But perhaps this approach is based upon a flawed assumption: that consciousness is an actual "thing" or even a process that is somehow "generated," like electricity or saliva, through physical processes that originate, most likely, in our brains. It is one where consciousness bridges a gap between an (imaginary) "inside I" and both inside and outside worlds.

But is that what really takes place? We might conclude otherwise: that our bodies or brains generate *ideas and concepts—including the notion of consciousness itself.*

Searching for Something Elusive

Considering these factors, it's not surprising that some scientists are still scratching their heads about consciousness. Perhaps some of them are

---

* Some Western thinkers such as David Hume were exceptions.

stumped because *they are looking for something that does not exist—at least not in the way they tend to believe.*

How We Might Come to Imagine Consciousness

Imagine watching a child study her reflection in a mirror and then touch her nose. We might think that, *through the lens of consciousness*, she is aware of her image and sees herself as an independent entity with an "I" inside, separate from everything else.

However, as we have discovered, that might not be what is truly happening. Instead, we might conclude that she *sees* the image; *thinks* of herself as separate from the rest of the universe; *has* the aha thought, "That's me!"; and touches her nose, driven by her curiosity to see the mirrored action.

Note the key distinction here. In the first approach, we infer that the child puts together conceptual pieces to accurately perceive reality. Yet, that seemingly accurate perception might only be a story. It is possibly a sibling to the debunked notion that, when we look at a green leaf, our brain captures an image of the leaf and then projects it onto its own mental screen.

Much of our exploration so far suggests that the second rendering of events is more accurate. We already know that the child is not a separate entity from the rest of the universe. There's also no internal "I," and her explanation of why she touched her nose may be a complex after-the-fact left-brain fabrication.

Think about Anna being hypnotized. When "Sarah" emerged, it appeared as if Sarah possessed consciousness. But if Sarah is only a figment or model, how can Sarah have "awareness," or realization, of anything?

All these observations suggest that we might think of consciousness as not involving a process of *realization* at all, but one of *creating* categories, concepts, and models—one of them being the idea of consciousness itself.

Through our journey so far, we have often encountered compelling, yet fanciful, stories of our brains' interpretive processes. These stories bring a consistent story to our lives. Yet, as we have seen, these stories often stray from reality.*

---

* Before continuing, readers might want to pause with this idea to reflect on how this paradigm shift might have an impact on the concepts/categories they have previously adopted.

Remodeling Consciousness

Questioning the Need for a Model

Given all this, we might question the need to model consciousness at all. Indeed, there is a need—for the same reason that we talk about sunsets and dial our smart phones. There is profound value in using concepts that we know are inaccurate, so long as we know that they are fabrications rather than accurate reflections of the self and consciousness.

We might therefore more accurately see that many of our concepts of consciousness depend on erroneously *thinking* that there is an observant "I" sensing the world.

This model both acknowledges a moment-by-moment sense of awareness *and* acknowledges the fabricated narrative that we habitually attach to it.

To show this point, let's revisit the story of Anna and Sarah. Anna was seemingly "aware" of Sarah as a second personality. Sarah supposedly had different memories and insights than Anna did. We can say that Anna was aware of this second personality and of her surroundings when she walked down those imaginary stairs. She might *think* that she perceived each of these things, both inside and outside of "her," just as someone else might be aware of a sunset or a phone call. Yet we also know that Sarah was just a character—a model—that Anna made up. The same is true of the staircase and the room at the bottom of those stairs. I believe that *the same is also likely true of her own "observant I," her sense of selfhood, and consciousness.*

This is not as far-fetched as it may at first sound. Science writer and Zen teacher Steve Hagen[3] frames consciousness as "the apparent dividing of a seamless Whole into 'the world we see'—i.e., the everyday world of 'this's' and 'that's.'" This might accurately describe our *perception* of self and consciousness, while avoiding the traps into which some other models of consciousness fall.

Hagen's insights also point to a prevalent flaw in numerous models of the human mind. We slice ourselves up into multiple personalities; into conscious and unconscious selves; into bodies with selves inside them; into "authentic" decisions and those made for us by demons, spirits, and other "outside" entities; into self and other; and into us and the rest of the universe.

In essence, what we often refer to as *consciousness* might therefore be thought of as a phantom bridge connecting nonexistent parts of imaginary dualities. By recognizing this, we might question the common assumption that consciousness truly lies within our bodies or brains.

Consciousness and Abstract Expression

Much like consciousness, abstract art can be perplexing. Some feel compelled to interpret it as something familiar, as if there is an apple or a person hidden in the brushstrokes. They might wonder, why did Picasso paint some people as if they were disfigured?

Gaining insight into abstract art can be a bit like unraveling consciousness. Both are chances to toss out our rule books and embrace the unknown. This is because how we think of both can challenge our conventional notions of reality.

With a piece of abstract art, there's a more rewarding way. That is to accept it as a unique creation that defies traditional categorization. The thing you are seeing in the art gallery doesn't have to be a person or an apple. Consider an abstract work like Wassily Kandinsky's *Composition VII*. At first, it seems to be a chaotic collection of shapes and colors. But as you immerse yourself in it, you soon appreciate the emotions and energy it conveys. It is unique and does not fit into your earlier "thing" categories. By seeing this, we can abandon our natural inclination to pigeonhole the world, including art and the idea of consciousness, into oversimplified preconceived molds. It is an exercise in flexibility.

In a similar vein, many of us think about "consciousness" in a commonsense way, i.e., as a noun. Here, it is seen as a "thing." But instead, we have learned that the *idea* of consciousness might better be thought of as a reflection of how we categorize and shape our reality. *Think of it as your brain creating concepts, just like an artist creating a masterpiece on a blank canvas.*

Such shifts in perspective become mirrors reflecting our willingness and capacity for reflection. They prompt us to question our beliefs, biases, and preconceived notions. Many nudge us toward growth and understanding versus self-awareness. They become paths toward personal growth and a more inclusive perspective on the complexities of life.

With such transformative shifts in perspective, we see that using *consciousness* as a noun or to refer to a unique process can prevent us from understanding the phenomenon itself. *Likewise, labeling an abstract artwork a "tree" prevents many from grasping the art's essence.* Thus, just as with consciousness, the creation and interpretation of abstract art is an ongoing and evolving process. Seeing this can both offer insights into ourselves and help us move past the confines of more simplistic and accepted understandings.

The Conscious Self as Concept

When we observe someone react to their environment, converse with others, or report their thoughts to us, we infer that they are conscious and have a sense of self—even if that conscious self is a fabrication. But perhaps that sense is not always hardwired into us. Maybe it's optional. Maybe it's learned.

According to reports,[4] a young Helen Keller (who was both blind and deaf) did not develop a sense of a conscious self until she mastered Braille and cultivated a relationship with her teacher, Anne Sullivan. It was through this journey that she learned that selfhood appeared to have been something she was *taught*. She wrote:

> When I learned the meaning of "I" and "me" and found that I was something, I began to think. Then consciousness first existed for me. Thus, it was not the sense of touch that brought me knowledge. It was the awakening of my soul that first rendered my senses their value, their cognizance of objects, names, qualities, and properties. Thought made me conscious of love, joy, and all the emotions.[5]

Could there be a clearer example of how consciousness and selfhood are purely models—or of how valuable these models can be, despite their illusory foundations?

Keller's situation suggests an intriguing thought experiment: What if her mentor had taught her not that she was separate from the world but an expression of it? How might she then have modeled the world, consciousness, and a sense of self? Or would she have conceived of them at all?

Despite her disabilities, Keller's story was not an isolated phenomenon. So here is another thought experiment.

We were *all* taught about selfhood, separations, consciousness, and "the external world" when we were young. For those of us brought up in Western traditions, I wonder what if, instead of the usual guidance, a philosopher following some Eastern traditions had shaped our understanding?

Enlightenment: Journeys beyond Models

The reality we can put into words is never reality itself. —Physicist Werner Heisenberg

Some Eastern philosophers often speak of *enlightenment*, which they say occurs when people suspend their customary habits of partitioning and analyzing themselves and the world. They drop their models—including the most basic models of self and "I" versus the rest of the universe.

Enlightenment is commonly described as a state of inner peace, bliss, even awe, and/or of being fully awake. While in this state, or when reflecting on it afterward, people often describe an experience of the oneness of the universe. They also often report feeling as if their sense of "I" or "self" dissolves.

Perhaps most notably, this feeling is linked with the abandonment of our adaptive, commonsense models that require separating ourselves from everything else in the world. We might describe this shift as leaving conceptions behind, or, as you probably remember, *no-model thinking*.

Recent research hints that this experience occurs when the left brain, known for its constant labeling, modeling, conceptualizing, and storytelling, takes a rest. When this happens, perhaps the right brain, which perceives the world in a more holistic, impressionistic way, takes center stage. With this, we might believe that we experience reality directly, bypassing the limiting effects of our mental models.

Yet even talking about enlightenment is fundamentally problematic. The feeling of enlightenment cannot be accurately described because no words can fully express—and no model can capture—a no-model world. Phrases such as "I experienced insight," or "I was enlightened," or even "There was no separation between my 'self' and the universe" still falsely imply someone who experiences something.

We thus see once again that words sometimes fall short in expressing profound concepts. The ancient Chinese concept of the Dao conveys this very notion. The Dao is thought of as the core mystery of life and something that defies verbal description. The classic text about it, the Dao De Jing,[6] even suggests that you are not really understanding the Dao if you attempt to articulate it. Once you define it, you have lost its essence.

In her journey of self-discovery, Anna encountered something similar. Psychological jargon such as "emotional eating" might have been used, but it could only capture small fragments of her complex difficulties. It could never convey all her ups, downs, and other emotions. Even the voice of "Sarah" helped broaden her perspective, although it was still incomplete. So, while these concepts and words helped, they still had their limits.

Thus, the Dao reminds us to see that descriptions can't capture complex realities. It might be better to experience something firsthand rather than to express, and confine, it with words or ideas. Then, concepts and words can be useful road signs but not fully capture the essence of things.

And so perhaps describing enlightenment, as well as some other truths, is always beyond our capabilities. But that's okay, especially as we come to realize, and accept, that not all things can neatly fit into our conventional worldviews.

This naturally leads us to another important question: How might we more fully model a human being?

Clearly, our commonsense model, reflected in common language, of *a "self" that is conscious of the rest of the universe* doesn't quite hit the mark.

The Fluid Self and Inner Peace

For most *utilitarian* purposes, I gravitate toward this model of humans: *A human is essentially a sentient being resulting from an unfolding interactive process within, and around, a* Homo sapiens *body*. Of course, there are endless ways that we might usefully conceive of humans.

One benefit of this model is that it could result in less suffering. At the very least, if we see ourselves as an ever-emergent and interactive process—and our sense of "I" as nothing more than a useful fiction—then change might not be so threatening.

By seeing the self as something fluid, more like a river than a rock, some people become less likely to anchor their identities on impermanent things. However, when people see themselves as static, basically as forever the same people stuck in the same bodies, they can start to blend who they are with what they own. Their possessions seem to become part of who they are. If you have lost something you treasured, you know that this has a price. You might have felt deep pain when you lost your things, or even when they changed. Your self-worth was wrapped up in stuff.

But those who can see themselves as ever-changing might also be less likely to tether their identities to mere objects. So, when they lose those things, it can be bad, but probably won't be devastating. They have learned to enjoy possessions without letting them dictate their identity. Simply, this view takes the edge off the ups and downs that come with losing material things.

This shift in perspective offers more than just freedom from material attachments. It also gives people the flexibility to endure life's curveballs, as well as a greater sense of inner peace. Some liken this to learning to surf where, with experience, they stop battling the waves and instead ride with them. They are agile enough to cope with changes rather than wasting energy resisting what nature throws their way.

This view could also lessen our fear of death. First, why does death frighten us? Well, it can be seen as the most drastic form of change. Here, it is seen as final, a permanent transition. And when we stick with a static idea of self, this can be terrifying. But if your identity is always changing, then "you" today is not the "you" of yesterday. You are truly having both "deaths" and "rebirths" every day. Thus, the fear of self-annihilation lessens, becoming part of the natural progression of life. The change we used to dread morphs into a realization of the beauty of life's ever-changing nature.*

After discarding the illusion of permanence, people can feel both liberated and humble. First, let's look at how it's liberating. So often we trudge along carrying the weight of our past, as if we had a sack full of stones. Each rock is a mistake or failure. But when you view yourself as evolving, your past mistakes might not matter so much to you. From this perspective, they might only be lessons. You are no longer tethered to your past, and thus find freedom.

But this view can also dish out some humble pie as you realize that your past good deeds and strengths were only temporary. Your newfound fluidity reminds you that the good person you were is not set in stone. Seeing this, however, might nudge you to always put in effort to be good, rather than resting on your past laurels. There is no time for complacency.

Overall, this mix of liberation and humility can lead us to a more balanced life. We are less weighed down by our past, and we understand the importance of consistently cultivating our virtues and never taking our strengths for granted.

---

* A psychological idea called *terror management theory* suggests that our fear of death determines much of what we do. It strives to ease people's fears that they will live insignificant lives and then pass away. Perhaps the attitude is like the common saying that "Life sucks, and then you die." Often people cope with this by trying to leave legacies, avoiding thoughts about death, or believing more strongly in an afterlife.

Consciousness Across Cultures

Like language, our ideas of self and consciousness evolve with societal changes. Tribal shamans and later medieval thinkers often held vastly different perspectives on consciousness compared to those of modern scientists. Thus, each culture shaped their unique understanding of consciousness.

These changing ideas guide the questions asked by both lay thinkers and scientists. In societies that prize individualism, thoughts often revolve around the self ("me, me, me") and emphasize free will. In contrast, other societies, where people greatly value collective cooperation, might stress the interconnections between individuals.

Our discussions about both self and consciousness should extend beyond the realm of science. We also must dive into culture, philosophy, and even our community's stories about who we are.[7]

Let's start with how some tribal shamans see things. For them, the idea of consciousness might be closely connected to nature, community rituals, and maybe even the concept of spirits. From their view, the "self" might not be an isolated thing in people's heads. Instead, it might be part of a larger natural world, their community, and perhaps even revered ancestors.

But in medieval Europe, the church influenced most parts of life. To it, consciousness was, and still is, often seen as a divine gift. "Conscience," for example, was intertwined with "consciousness" and was viewed as a gift from God to help people know right from wrong. Thus, lapses in morality could be viewed as neglecting your God-given conscience. For instance, if Anna commonly overate, her weight problem could be considered a sign of gluttony, one of the deadly sins.

Modern Western societies highly prize individualism. Thoughts about "consciousness" often focus on free will and an individualistic "It's all about me" identity. Unlike in many other parts of the world, this Western view places less value on community, teamwork, and interconnectedness.

Perhaps it's time for a shift in perspective. What if we turned our focus from "What is consciousness?" to "What notions of consciousness would make the world better?" And so, our discussions about consciousness aren't mere academic banter. Instead, they become tools for reimagining society. They challenge us all to think beyond ourselves and perhaps adopt a more detailed, perhaps even more compassionate, approach to life.

CHAPTER EIGHTEEN

GENEROSITY BEYOND GENETICS

Throughout our journey, we've explored different models of human thought, often borrowing insights from both Eastern and Western philosophies. These traditions show that, although what we believe is sometimes wrong, such mistaken ideas can still serve us. Take the common heartwarming belief that humans are, deep down, selfless. It's a comforting idea but one that, I'm afraid, causes a lot of debate.

Carl's Path to Connected Living

Now, allow me to introduce you to Carl. He was a 35-year-old patent attorney whose life in Chicago was anything but serene. His purposes have always been about the next case, his next win, networking, and promotions. He never put thinking about others' happiness high on his priority list.

Amid his hectic courtroom battles and high-stakes negotiations, he became almost crippled by severe panic attacks. His heart beat uncontrollably, he breathed hard, and sometimes he felt as if he would lose consciousness or even die. His attacks would ambush him at the worst times. He was usually in the spotlight when they happened, something that was inevitable in his high-stakes job. All he often wanted to do was escape to the comfort of his home.

The money was great. But he soon knew he couldn't continue like this. Determined to find a solution, Carl decided to see a therapist. The therapist floated an interesting idea: "What about volunteering? Perhaps it would distract you and help you gradually feel more comfortable in low-stress social situations. After some practice, you might later be able to return to other more high-pressure work situations but have less distress."

Let's zoom out to better appreciate the reasoning behind her advice. Giving others a hand is a big part of who many of us are, even if Carl didn't do that often. We see this with empathy and compassion. Empathy allows us to sense, understand, sympathize with, and connect with others' emotions. Most of us feel this at least occasionally, and sometimes it even helps with our own struggles.

Let's get back to Carl's story. He took his therapist's advice by volunteering at an animal shelter. At first he was terrified about going into the facility. It was nerve wracking. His almost constant worries made him feel caged, and he could barely socialize with other staff.

However, armed with an attorney's tenacity, Carl slowly began to have a transformation. One of the first signs of this change was his shift in perspective, now often focusing beyond himself. He realized that even while grappling with his panic attacks, he could still help some of his furry buddies to find homes.

The more he helped the animals, the more his anxiety receded. It felt as if his brain finally switched tracks, with his kindness becoming an integral part of his therapy. Not only was he helping things beyond himself, but he was creating a more peaceful mental state within himself.

It's fascinating that Carl's experience points toward something universal and not just sentimental idealism. This is because studies show that compassion and empathy change brain chemistry.[1] For instance, after doing something nice for someone else, our brains can release more of the "feel good" hormone oxytocin, something that often lowers anxiety. But it also strengthens friendships and even makes romantic feelings stronger.

Let's dive a bit more deeply into anatomy by talking about the "mirror neurons" in our brains. These fire up when we watch others do certain things.[2] For example, if you see someone fall, you wince and think, "That must've hurt." That is likely your brain's mirror neurons chiming in, helping you share someone else's experiences.

Carl's progress showed this principle. His initial self-centered search for tranquility changed. Eventually it evolved into genuine empathy and a desire to care for something beyond himself. His acts of kindness boomeranged back, reducing his anxiety, and fostered emotional connections with the world.

Carl's improvement fit well with the Eastern idea of "interbeing." This stresses that all life is interconnected. It says that our own actions influence both ourselves and the world. We saw this in Carl, when he eventually had more desire to care for something beyond himself. His kind acts, in a way, came back to help him lower his anxiety and feel new emotional connections with the world.

Now let's broaden our view a bit more. Most cultures hold a special place for the spirit of giving. In my experience, it is something that often resonates with secular and religious communities. Each of these often encourages people to transcend their egos and instead put themselves aside for the greater good. These communities truly value compassion and empathy. They also appreciate how all beings are interconnected. Most of us, of course, hope that all of this is an important part of human nature.

Unpacking True Altruism

The term *true altruism* captures the notion of being selfless without expecting anything in return. This perhaps idealistic concept often stirs up passionate debate about what it means to be human. Skeptics abound, often doubting that true altruism really happens. Some studies on our brains' mechanics support this idea. These reveal that our brains tend to reward compassionate deeds with a delightful chemical high. They have found payoffs in what can otherwise appear to be true selflessness. This discovery logically blurs the line between selflessness and self-interest.

Knowing this, it's not hard to question our motives. Are they driven by true altruism or by the anticipation of a payoff, in this case a wonderful feeling brought about by brain chemistry?

Evolutionary biologist Richard Dawkins[3] has strong feelings about this. He argues that our genes lean toward selfishness and questions whether altruism fits with the principles of natural selection. In his own words:

> A predominant quality to be expected in a successful gene is ruthless self-ishness. This gene selfishness will usually give rise to selfishness in individual behavior. . . . Much as we might wish to believe otherwise, universal love and welfare of the species as a whole are concepts that simply do not make evolutionary sense.

Picture giving your kids a nice financial boost, helping your sister turn her house into her dream home, or letting your cousins crash at your place while they are hunting for jobs. Each of these actions, of course, can *bring you* a sense of fulfillment.

But from Dr. Dawkins's evolutionary standpoint, such "sacrifices" aren't just simple kindness. They make us feel good, sure, but they also come with other benefits. This can happen when we help our relatives, because helping them can improve the chances that our shared genes will survive. By assisting them, Dawkins might therefore say that you are really investing in *your own* genetic future.

A similar rationale applies when we seemingly sacrifice ourselves for nonrelatives. Some interpret these actions as only one more route to genetic success. Here, cooperation and supporting friends aren't about giving. Instead, they are seen as strategic moves that will later benefit everyone by helping entire societies flourish.

Dawkins's views fit well with individualistic models that see humans as separate from, and in competition with, the world. These have been important concepts in many Western cultures that emphasize self-reliance and personal independence.

And so, if Dawkins's thinking is correct, we see then that what masquerades as selfless giving is instead often about safeguarding our own genetic legacies. Obviously, that would not be what many commonly think of as altruism.

Beyond Cynicism

Some argue that it is a grave error to dismiss true altruism. They point out that actions can sometimes be altruistic if there is no obvious payoff for them in return. A few people even worry that Dawkins's outlook could hurt society, as they say it promotes selfishness and erodes moral values.

These dissenting voices passionately assert that true altruism not only exists but is as natural and common to us as eating and breathing. They often, in fact, believe that it can change the world.

Finding a Middle Path

Despite this emotional debate, we can achieve a consensus to bridge the divide: aspire to put the greater good ahead of personal gain. Helping others should not only be about genetic continuation but should also foster community bonds and mutual support.

With this attitude, we can still use Dawkins's theories to interpret nature, rather than as behavioral expectations for civilized societies. While he zeroes in on the mechanical, we thus can focus on goals for better communities.

Moreover, here's an uplifting thought, even for those who don't buy the idea of true altruism. Research[4] suggests that if we encourage some people to empathize and think about how others feel, they become even more likely to help. This, as we are likely to guess by now, happens partly because doing so dials down *helpers'* discomfort. That's just fine with me. Consequently, the debate of true altruism might be less important than just doing the right thing.

Beyond the Semantics of Selflessness

I sometimes wonder if all the bickering over the existence of true altruism isn't just a tussle over semantics. If we consider giving actions (even those followed by chemical rewards) as being altruistic, this dispute fades. The focus then is turned to the kindness itself.

With this shift, we would also accept that receiving a chemical boost doesn't diminish our deeds or our moral character. The positive result is the most important thing, even if we don't think the motivation is noble. In each case, our biology predisposes us to actions that align with moral and social aspirations, whether motivated by chemical rewards or not.

In the end, despite our views on this, it's crucial that we not become cynics. Even if our thoughts about altruism shift, or even sour, this doesn't have to imply that we're all selfish and only pretending to be nice. Much as Anna's unfounded memories led to positive change, our own capacity for compassion and generosity, for whatever reason, remains a powerful force for good in our world.

The Dangerous Terrain of Self-Righteousness

Strongly believing that we are altruistic, or even just believing in our own high moral character, can have a downside: fostering the fantasy of moral superiority. Basically, if we think we are above selfishness, we risk becoming complacent. This can also lead to "moral licensing,"[5] where we excuse selfish behavior because we think we've already "done our good deed for the day." For instance, if you just volunteered at a homeless shelter, you might then conclude that it's fine to turn away from someone else in need.

This has tangible consequences. Some feel entitled to be less environmentally friendly if they, or their peers, recently behaved morally. Recognizing this tendency can help us stay grounded. Good deeds are great, but they are no free pass for other ethical lapses.

Altruism Masking Autocracy

Believing in unrewarded altruism, which is perhaps part of being too trusting, can also backfire. Take charismatic leaders, for example. They are often masters at painting themselves as self-sacrificing altruistic champions of the downtrodden. These lies can lead their followers to see them as virtually perfect. It is as if some people are seeking a messiah, a person to fix all their problems. Guess what happens after this? Nobody questions the leader. What better way is there to get an autocratic regime up and running than by unwavering obedience to a leader who only claims to prioritize the public good?

Cultivating Altruism through Collective Well-Being

Skeptics of an altruistic nature often wonder how humanity can transcend its selfish instincts. Various philosophical teachings suggest that we could do this by cultivating a spirit of self-sacrifice.[6] For instance, when we encourage people to support worthy causes, this can amplify their natural compassion. And so, one key perhaps lies in seeing how one's own welfare is intertwined with that of others. This type of understanding sometimes inspires more collective approaches to altruism. It is one where the line between helping oneself and helping others is blurred.

Such blurring would obviously echo the philosophical idea of "no-self," where the self is not a separate entity. By seeing this interdependence of all people, a new path becomes visible. *It is one where altruism becomes a natural expression of the inherent connectedness in our lives.*

We have now seen that even if we think that pure self-sacrifice is hard to find, it doesn't mean we should abandon our efforts to be compassionate. Instead, we can encourage empathy and generosity to spark change. We might also remind people that we're all in this together. Each of us knows only a collective journey. This echoes President Kennedy's sentiment when he said "a rising tide lifts all boats." This concept still applies today.

CHAPTER NINETEEN

SELF, CULTURE, ARTIFICIAL INTELLIGENCE, AND PHILOSOPHY

In the 21st century, we have seen technology race ahead like never before. It has often improved our lives along the way. Artificial intelligence (AI), for instance, is now a catalyst for fresh and diverse opportunities, some of which have already led to both prosperity and personal fulfillment. Communities across the globe have reaped its benefits, including enhanced medical diagnostics and surgery, improved business practices, more accurate predictions of wildfires, and advances in the fight against climate change.

Think also of the personal benefits your computer AI assistant might already have brought you. Perhaps yours, like mine, is named Cathy. She not only helps you manage both your schedule and your budget, but just before bedtime she even reminds you to let out the dog. She seems genuinely interested in your life. Although you might be grateful, you could also wonder, "Was I helped by a tool, or by a true companion?"

AI, of course, has many dangers. Terrorists can use it to make weapons, conduct cyberattacks, and even wage biological warfare. Another less severe but nevertheless justified concern is job loss. In response, screenwriters have successfully curbed AI's influence on script writing. Think also about AI programs that don't just research lawsuits. They are also "smart enough" to suggest courtroom strategies, reducing legal counsel's billable hours. In the face of such changes, even lawyers have something to fear.

Some believe that AI is from the Devil. Others worry that AI can be a threat to their autonomy. Current AI technologies, for instance, already encourage us to eat certain foods, direct our search-engine findings, steer us to certain products, and anticipate the music that we like to hear.

Perhaps we can mitigate some of these problems with morality and a look at our philosophies. For instance, attitudes that value our collective well-being through environmental sustainability, rather than personal gain, might help steer AI to altruistic purposes.

But to help people accept, adopt, and potentially even use AI for purposes that benefit us all, let's remember that there is a lot of work ahead of us. It involves shaping attitudes toward this technology. These attitudes, of course, are not formed in a vacuum. Rather, they are influenced by such factors as location, financial status, and education.

For instance, people in tech-savvy metropolises, such as Tokyo and Silicon Valley, already tend to see AI favorably. Those with more education and financial clout also are often inclined to view it positively. On the flip side, some people with fewer resources might have more concerns that AI will widen the gap between the haves and have-nots.

Aspirations

Given many people's apprehensions, perhaps it would be wise to connect millennia-old thinking, which has successfully guided humanity in the past, with AI. Our resilient ancestors, who adapted to many changes in the environment, developed dynamic, flexible survival strategies. My hope is that, partly by encouraging this mindset, people will be more willing to accept modern innovations, including AI.

Idealistically, I imagine the world this could create. My vision is of societies in which our evolving selves and groundbreaking technologies are seldom at odds. Instead, using the kind of flexibility passed down from our ancestors, along with human modesty, curiosity, and the capacity to engage with the abstract, they would be collaborators in progress, each learning and growing with the other. These are, of course, the qualities and attitudes promoted throughout this book.

More Philosophical Reflections in the Age of AI

Currently, researchers and therapists who embrace the concept of "transhumanism"* use brain-computer interfaces, either with surgical brain implants or electrodes attached to the scalp, that improve humans' natural abilities. Therapists use them to accomplish such goals as helping patients move robotic limbs and computer mice as well as to improve vision. We might therefore ask, "Where exactly does the person end and the computer begin?"

---

* This is the idea that we should use available technology to enhance human function.

Such philosophical questions invite yet more thought about AI. One of them pertains to some people's model that humans are an inherent part of everything that exists, rather than self-contained units. We might wonder if those following this idea might be just the people most likely to eventually view AI as part of their personal journey. They might even come to see AI as an extension of their "self."

How could this happen? Well, let's think, for instance, about selving, a concept we explored earlier. It offers parallel examples, such as when people often integrate not only other humans but sometimes material possessions into their own sense of self. Just as fancy cars and romantic partners can literally be seen as an extension of one's identity, so can our interactions with AI lead to a conceptual merging between humans and computers.

In fact, the boundaries between the self and outside, thus softened by AI, already help us see ourselves more broadly. It is a vision of humans as both interconnected and evolving with AI.

Researchers who embrace this attitude, clearly reminiscent of some Eastern thought, argue:

> Connecting our brains directly to technology may ultimately be a natural progression of how humans have augmented themselves with technology over the ages, from using wheels to overcome our bipedal limitations to making notations on clay tablets and paper to augment our memories.[1]

Therefore, it's now not too far-fetched to see how many people welcome AI, *even modeling it as an extension of themselves.* Adopting this new model of "self" might, in fact, be the next logical step for more of us. From this idealistic angle, AI would not be an adversary but rather an ally.

Harmonizing AI, Culture, and Change

Some people are unfazed by AI. Maybe they don't keep up with warnings on the news or don't think it will personally affect them. Another possibility, however, is that they are generally the ones most likely to accept change.

Perhaps such changes will roll in sooner than we think. Yet, if we are wary of the unfamiliar, of novelty, or of shaking things up, we might instead miss out on the promises AI offers.

Time-honored wisdom, like that found in the I Ching[2] (often translated as "The Book of Changes"), has already helped some embrace change brought about by AI. Of course, this new technology is only one example of change, the universe's only constant.[3]

Then there's humility. The more of it we have, the fewer concerns some people might have about AI. This is perhaps because humble people don't find it narcissistically excruciating when they no longer feel as if they are the best at everything. A more realistic modesty might therefore help them feel as if AI is a partner in progress rather than an adversary.

Humility in relation to AI might be an especially worrisome problem in the West, where we strongly emphasize both individuality and human superiority.[4] Remember when, for example, humanity learned that the Earth was not the center of the universe? That knowledge threatened the idea that humans were superior to nature rather than only part of it. Well, perhaps something like that is going on now, as humans first encounter machines that are, in some respects, "smarter" than themselves.

Beyond Fear: Our Future with Technology

It's surprising that some ancient traditions do not see modern technological innovations as so much of a threat.[5] For instance, a Pew Research Center study found that more than 65% of the residents of some Asian countries believed that AI is good for societies, compared to 47% in the United States and only 37% in France.[6] Rather, some in the East even welcome AI as part of the natural course of human evolution. AI is, in fact, becoming a more comfortable part of their existence.

This might not be surprising, as some Eastern societies have incorporated mechanization into their spiritual beliefs for centuries. Think about ancient Tibetan prayer wheels, for example.[7] These cylinders, which are spun clockwise, have sacred symbols on the outside. The devout often believe that this spinning motion gives the same benefits as a human reciting the sounds, words, or phrases contained in the symbols. More recently in the East, there are digital prayer beads, as well as solar-powered gravestones that chant mantras.[8]

Some Eastern countries have been especially welcoming of mechanization because of the scarcity of human workers. Japan, for instance, has shown a decades-long trend to favor automation over manual labor. Its culture thus has widely embraced robots for both industrial and other uses.[9] Thus, certain cultures are primed to adapt technologies such as AI, even for the most personal needs.

For instance, visualize going to the tranquil Buddhist sanctuary Longquan Temple in Beijing, China. When you arrive, you are greeted by a blend of spirituality and technology. It's a plastic, advice-giving, bilingual, AI-equipped robot monk, scooting around on wheels while dressed in traditional monastic clothes.

Xian'er the robot, used there for about a decade, was designed both to answer questions and to chant Buddhist mantras. Some joke that it is often used to reach people who are just a little too attached to their smart phones. Xian'er is so popular that he has been seen by millions of viewers on social media (including Facebook) and appears in comics.

This robot already nicely harmonizes ancient traditions with modern technology. However, its abilities are limited. Xian'er is not, for example, yet able to solve complex problems.

Pointing out similar difficulties, one author said that many "have questioned whether or not robots can believe in the ideas they espouse, whether or not they have the emotional capacity to mediate meaningful interpersonal relations, or whether or not they have sufficient moral capacity to make ethical judgements."[10]

In all, such practices show AI's capabilities. They do, however, also raise questions about its limitations.

Clearly, the benefits we might realize from AI depend on our willingness to accept and ethically use the technology. They also are dependent on our own flexibility, willingness to embrace diverse philosophies, and morals.

And so, to best navigate the new age of AI, let's be open to change. Part of this will involve always critically, and ethically, examining our evolving relationship with technology. It is only with such practices that we might gain the best chance of unlocking its potentially transformative benefits.

Part III

WHAT WE ACTUALLY SEE

When I point out that Mind is grasses and trees . . . it startles your ears.

—Dogen Kigen

CHAPTER TWENTY

EMBRACING A UNIFIED REALITY

We have learned that nearly all our models—and all our conceptual divisions—are at best useful. They help us navigate our lives but do not accurately describe or explain the unified reality of all things. They are like pixels, algorithms, and digital code: At their best, they simulate or approximate reality, but they can never fully reflect it.

More notably, when we mistake those simulations or approximations for reality, we impede a full understanding of ourselves and the world.

Understanding this, physicist Roger Jones wisely cautioned,

> When we treat our representations [models] as reality, when we forget their original purpose of merely saving the appearances, we commit a profound kind of idolatry. We forget that we created our physical representations and metaphors, and we imbue them with a power and independence that ultimately comes to intimidate and control us.[1]

His advice underscores the profoundly misleading impact our conceptual frameworks have on our understanding of the world. For instance, we know that, according to most commonsense models, our minds are somehow separate from the rest of the universe, lodged inside our skulls. Nevertheless, these minds are utterly unlocatable. And, somehow, these minds are often thought to operate independently of their surroundings because of free will.

By now we have learned that these models have limits.

So, what if we were to discard all these artificial divisions that contribute to this limited view of mind? This would be more than an exercise in philosophy. It would be an opportunity to reassess our basic assumptions about humans and explore new ways of thinking.

By doing this, we might view mind as *being* all things—including social contexts and relationships; the interactivity of all action, speech, emotions,

177

impulses, and thoughts; and even the parade of faux specifics that we call hypnosis, unconscious motives, evil spirits, and so on. From this perspective, we would acknowledge that an "enskulled"[2] mind is mythical and cannot be located inside our heads; indeed, it cannot be found in any *one* physical location. We would accept the view that, in essence, *mind is everywhere*.

Similarly, what we call *hypnosis* is also everywhere because it is part of this interactivity. It includes all that surrounds us and is in us. Seeing this, *we realize that hypnosis is emblematic of how humans are interconnected with all else*. This is a far cry from viewing hypnosis, and ourselves, as centered in the brain. Seeing this is to look at ourselves through a lens of interconnectivity.

We find that if we welcome the idea of mind as all-encompassing, extending far beyond the confines of our skulls, we profoundly shift our perspective on ourselves. This broad view, where "mind" is intertwined with all human interactions, brings an important acknowledgment: *Temporary things that we call hypnotic trances, malevolent deities, and unconscious motives are not limited to our heads but are instead part of a far larger interconnected and always-changing world.*

When understanding, and experiencing, their own inherent bond with the world, some are struck by a profound sense of awe. *Connecting with the world like this helps them grasp their own part in something incredibly vast and wondrous.* This perspective is one where the boundaries between themselves and the universe are blurred. It turns the commonplace into the extraordinary.

Useful Fabrications

We have learned that our reasoning processes fabricate things (like ids and evil spirits), processes (such as unique hypnotic inductions and trances), explanations ("A little girl slipped her arm into my sleeve"), and motivations ("I wanted to get a Coke" or "I pressed the button because I decided to do so"). It likely also extends common concepts of consciousness in relation to an "I" that only "takes in" what is around.

All evidence suggests that our brains also manufacture the very concepts of space and time. This helps us thrive. However, they also create metaphorical boxes that limit our perspectives. By appreciating this, some of us are tempted to get a broader glimpse of reality and step away from the comfort of past belief.

Thanks to Einstein and other physicists, we now know that space and time are not separate but are more accurately seen as a unity called space-time.* Thus, our concepts of space and time only *seem* to make sense within the confines of our own reasoning and perception. In fact, they are just models—two of the most common, among many.

What does all of this tell us about ourselves? That we are enormously flexible and adaptive—and that we have at our disposal an almost limitless supply of poorly supported, but sometimes useful, concepts and models.

Anna's "hypnotic induction" is an easy-to-grasp illustrative example. In this seemingly altered state, she "remembered" events from her childhood that never took place. However, those imagined states, and the false memories, ultimately strengthened her therapeutic relationship with her therapist and conceivably played a role in her recovery.

In summary, when it comes to many of our mental models, we appear to be wired with an affinity for "Hey, whatever works."

Beyond Separateness

As we've seen, nothing exists independently. Any apparent separateness is, at most and at best, a useful idea. It is a concept that sometimes lubricates human interactions—and may support our species' survival in other ways.

Common ideas about hypnosis can also be seen, in part, as useful models. So can an exorcism of an evil spirit, presided over by a religious leader. So can the denial of climate change models by a politician, as well as the acceptance of climate change science by their political rival. The success of each such model depends on its social and interpersonal contexts.

The impact of these influences often flows in more than one direction. To explore this, let's return to the example of hypnosis. We know that the personality, claims, ideas, and beliefs of the therapist who does the hypnotizing can have profound effects on the patient. But "hypnotic induction" is a shared interaction, *just like all our dealings in the world*. This means that the patient's personality, claims, ideas, and beliefs can also impact the therapist. Indeed, patients can cause hypnotists to mimic their patients' breathing patterns, develop greater empathy for them, and respond in other measurable ways. We can even say again that a patient could partly hypnotize the hypnotist. In such ways, hypnosis can be seen as an interactive social dance.†

---

* In 1240, the Eastern philosopher Dogen Kigen wrote about this unity (without the math) in his essay "Being-Time." Dogen has been described as a "mystical realist."

† This somewhat blurs the roles of therapist and patient. It challenges the typical concepts of power dynamics in therapeutic relationships. If both people influence each other, we

For example, a patient's sense of wonder, skepticism, gullibility, or confusion can change the hypnotist's view of therapy, science, philosophy, hypnosis, psychology, and even their own abilities. The patient's words and actions might (wittingly or unwittingly) nudge the therapist toward a different direction in their work, or their research, or even their conceptual models of the human mind. In this manner, a patient can become a lens through which the therapist learns to view the world.

If a therapist tells a patient to look for a second personality inside themselves, the patient will be tempted to find one. This happened with Anna as the therapist planted a seed of expectation, one that grew into her actual experience. We see in this example that our brain's ability to create experiences based on suggestions is stunning.

But we also know that this process goes both ways. The more patients discover and converse with such second personalities, the more therapists accept their existence. Patients' hypnotic behavior therefore helps guide the therapist's reality.

When a patient acts "abnormally" while in an apparent trance, this reinforces to the therapist that hypnosis is a special state of mind. However, the supposed "abnormality" of those actions (as well as the concept of hypnosis as a unique mind state) is sometimes a useful fabrication partly created in the therapist's mind.

All of this occurs within the context of evolutionary biology, which determines our perceptual abilities and predisposes us to form and embrace models that support our own best interests—i.e., our ability to survive and thrive as a species.

Finding Truth beyond Our Creations

The basic paradox with which we struggle is that our survival depends on our models, yet they are flawed. Their flaws, however, can show how to engage with the world in more meaningful ways. The idea of consciousness includes such useful but faulty divisions. These divisions are of literally anything we can think of, including ideas about space, time, and consciousness itself.

---

therefore place a bit less credence on the idea that a therapy relationship is solely to allow an expert to help a patient.

It's like a patchwork quilt. Each of the quilt's squares represents a model that organizes our experiences. But when we view it from afar, we see that each tiny square is part of a far more intricate design.

For decades, we have grappled with the question, *How does the brain create consciousness—i.e., a means for perceiving an outside world?* But we have discovered that this question cannot be answered in the way that most of us had hoped and expected.

We have found zero evidence that the human brain—or any brain—on its own somehow creates consciousness, in the sense that consciousness is a means to perceive a world beyond an inside "I." Instead, the *idea of* consciousness might stem from our endless conceptual differentiation—our endless model making—out of our universe.

Meanwhile, our brain's interpretive process fabricates one false but believable story after another to support the validity of our ideas. Thus, our brain accepts our stories and models as reality itself, rather than as useful fictions. Our brain misleads us about both our own nature and the world at large.

It is difficult for us to see this in ourselves. Nevertheless, as this book has hopefully shown, better awareness is possible. If you succeed, it's like taking off dark glasses you had forgotten you were wearing. Suddenly, the world looks brighter and more nuanced.

In this respect, I'm reminded of Atticus Finch, a character in the movie *To Kill a Mockingbird.*[3] Atticus is a lawyer in the American South defending a Black man falsely accused of raping a white woman. The movie plot addresses racial injustice and moral growth.

In one scene a rabid dog, suddenly threatening Atticus's community, appears. The dog reminds the viewer of such things as racism, prejudice, and moral decay.

Shortly before shooting the animal, Atticus removes his eyeglasses. With this, he symbolically gives himself a clearer and perhaps more detailed understanding of the situation. This act has, of course, much relevance to some philosophies. It could be seen as representing removing one's "conceptual lenses" to better fathom reality. The glasses, in this respect, can be viewed as filters. In the movie the filters signified limited views and prejudices, each of which helped shape the community.

There is also the closely associated Eastern concept of "Right View," which encourages knowledge not tied to what we want to see, or to our fears, but to how things truly are. Atticus's act expresses this same sentiment when he abandons his preconceptions (symbolized by his glasses) to confront the life-threatening situation at hand.

Further, Atticus's action is like some Eastern teachings of the always-changing nature of life. He was primarily a gentle man, very interested in scholarly pursuits. But when confronting the rabid animal, he shows the fluid nature of his "self." That is, he was willing to take on an act contrary to his usual nature to defend others from danger. Like Atticus, none of us should be permanently wedded to an unchanging way of being.

Finally, one can also sense compassion in his actions. He does not enjoy killing the animal yet does it to protect his family and community. It is a last resort that resonates with the idea of doing whatever causes the least harm. He has adopted a role that he would prefer to avoid, but one that is necessary for the greater good.

His actions, performed in a small town, point to common truths about ourselves. Unless we remove our metaphorical glasses and see our thoughts for what they are—imperfect models—they can lead to confusion.

Where does all of this leave us? We might feel helplessly adrift. How do we survive and thrive in a world that turns out to be far different from what we think it is? How do we process the awareness that we humans are not what we usually believe we are?

Seeing the truth about ourselves can be discomforting. But discomfort can often be good. It can signal that we're on the edge of new realizations, and perhaps ready to rethink our world. In fact, we'll see in the next chapter that it also offers us some profound opportunities for enhancing our lives.

CHAPTER TWENTY-ONE

TRUTH, HUMILITY, AND A WORLD BEYOND MODELS

Evolutionary biologist Richard Dawkins is simultaneously optimistic and apprehensive about our ability to improve our world. He observes:

> If any species in the history of life has the possibility of breaking away from short-term selfishness and of long-term planning for the distant future, it's our species. We are earth's last best hope, even if we are simultaneously the species most capable of destroying life on the planet.

His belief that we might put our immediate self-interest aside in favor of longer-term goals is uplifting. However, his thinking also points to our destructive potentials. We all hope that, by realizing this duality, people would keep close track of both their beliefs and actions.

The Mirage of False Models

Sadly, there is ample evidence that we do not have enough of this self-awareness to fulfill Dawkins's dream.

As we've seen, we often have far too little insight into the actual causes of what we do, decide, think, or conceive. We routinely ignore reality in favor of false but comforting, convincing-sounding, and often useful models and stories.

To the degree that these models and stories help us survive, both individually and as a species, we continue to have opportunities to plan for the distant future. Often, however, our convincing-sounding and widely accepted ideas are not only false but self-centered and ultimately destructive. They are like mirages in the desert. Some look convincing, but if you pursue them, you could be in serious trouble.

When we use any false model and then later realize that we have done so, we might say that our actions were an aberration—the result of an altered mental state such as hypnosis, demon possession, or even a nonexistent mental illness. We might thus say that we were influenced by a force stronger than ourselves. Too infrequently we say, "I was foolish and selfish. I caused harm or created problems. I'm sorry. To plan for the future, I'm going to investigate what would be wiser and more realistic."

Adaptive Mindsets and Ancient Wisdom

That said, our ability to create, adopt, and adapt a wide range of models is one of our species' greatest strengths. As an example, consider these observations from an eighth-grade science teacher:

> As a Christian, I believe we are called first to understand science because it explores all of creation, and by understanding creation we understand more about our Creator. In my own life, science is what inspired me to start asking questions about and engaging with the world around me, rather than simply going through the motions. I teach science because I want the next generation to be able to think critically and reason logically, to explore and discover more than we know today, and to find joy in observing deeply the world and the skies. This can ultimately lead to gratitude for and stewardship of the natural world.

Both God the Creator and Christianity can, of course, be seen as models. But acts of gratitude, curiosity, inquiry, critical thinking, and stewardship do not merely reflect how we look at things. They are also *behaviors* that inspire human beings to break away from short-term selfishness and plan for the distant future.

Faith and Inquiry

In this teacher, we see a caring person who models the world in terms of both religion and science. She continues to make a positive change in the world right now and shows a commitment to service and virtue. She does not let herself be diverted from these goals by focusing only on thoughts of rewards and punishments in an afterlife. She is guided both by the teachings of her religion and by a trust in science. This is an inspiring example of not only adopting and expressing multiple models at once, but of doing good things with each.

Just as importantly, by embracing science, this teacher knows that she must change her thinking as new facts become available. Certainly, in all areas of our lives, what we decide to believe, how we decide to help, and the methods in which we have faith must often change to be effective. This is because, in a shifting landscape, the methods in which we've trusted might soon no longer work.

Consider the advent of computers. In the 1960s, the best way to help someone was often to give them a book. Today, it's often far more helpful to provide access to the internet, and perhaps to an online course.

Furthermore, stubbornly insisting on following outdated, perhaps even literally ancient, thinking when we know it no longer helps others isn't just ineffectual but immoral.[1] It's like seeing a doctor who prescribes an outdated medicine when a newer, more effective one is available.

This adaptive mindset goes far beyond ways to help others. The way in which we perceive the world and ourselves, and unwavering beliefs in old principles, tether us to the past. In fact, rigidly adhering to any idea, without occasional reconsideration, can be a type of self-sabotage. It's like refusing to update the software on your computer. Soon it will no longer work, and you're sure to get a virus.

Social Diversity and Possessions

It's important to remember that much of the richness in our lives, and the strength of our communities, comes from diversity of thought. Our strength and joy lie not in uniformity but in the many perspectives, experiences, and philosophies each person contributes. When societies eagerly encourage such diverse thinking and fresh ideas, they consistently bring about innovation, creativity, and constructive change. Just as a forest having an abundance of diverse plants, soils, and animals thrives, so too does a society profit from the many worldviews its members explore and adopt. If we stifle this freedom, we also choke evolution and growth.

Another key principle in fostering productive change might be to deemphasize the importance of possessions. This is because relationships, experiences, and personal growth are the real treasures that enrich us. Physical items, however, are often burdens, as well as distractions from other people and our values.

This doesn't mean that Buddhist monks, who strongly value austerity, can't have the latest smart phone. Perhaps they view cell phones not as riches but as ways to better connect with others and spread their teachings. Seen this way, owning a telephone does not contradict their values but instead shows their flexibility in ways to spread knowledge. They have embraced technology and, by doing so, can better improve their worlds.

The issue, then, is not one of relinquishing almost everything, but of achieving balance and harmony. This accepts the complexity of our world. It also shows that the values of a nonmaterialistic life can coexist with modern technology. Both help us better engage with others.

Collective Prosperity versus Personal Gain

While we acknowledge the value of frugality, we can still feel the allure of self-centered craving. Desire, in fact, is a deeply rooted part of our nature. Some of us have truly insatiable desires for power and possessions. We have all seen times when this intense, self-centered focus on wealth destroys human relationships.

But there is hope for bypassing excessive materialism. We might, for example, educate people about both the fleeting nature of material possessions and the value of relationships. By further teaching others to cherish what they have and to see the transient nature of everything, the relentless allure of desire sometimes diminishes.

Encouraging cultures where collective growth trumps individual success can foster other changes. If we teach that success is not seen so much as a personal achievement but instead as a collective effort, we might lessen the pull of self-centered craving.

Finally, some of us find value in practices such as meditation. It can enhance our awareness of, and perhaps change, our destructive self-centered thoughts, desires, and actions.

Each of these efforts fits in well with a more universal framework[2] for understanding well-being. For instance, principles such as avoiding excessive desire-driven pleasure and improving moral conduct are found, in some manner, across most cultures. Each idea thus taps into foundational elements common around the globe. This means that most communities might be open to further embracing, developing, and sharing these ideas.[3]

Personal Identity in Virtual Worlds

Technologies, such as artificial intelligence and virtual reality (VR), have already changed how many people perceive themselves and reality. Some people have even seemed to find new identities through them. They have sometimes broken free from both physical reality and familiar social norms. In these digital worlds, they can have new selves. With this change, technology has provided playgrounds for developing new identities that are impossible in the default world.

This advancement in technology therefore blurs the boundaries between the virtual and the real. We perhaps wonder what will happen when it becomes even more complex, and perhaps eventually replicate the most private human experiences.

The topic of AI and VR changing our self-perceptions has evident parallels with Anna's journey. Recall Anna's self-examination through hypnosis, where she temporarily shifted from one person into two. In some ways, her metamorphosis was a needed change.

Anna's actions were therefore not only a way to adapt to an ever-changing world but show the sometimes *necessary* flexibility needed to bring goodness to ever-changing times. Such mindsets help us create new, more adaptive "selves" and flexible concepts of morality, each of which becomes newly equipped to generate solutions to previously nonexistent problems.

The skill of adapting our sense of self can be incredibly freeing. Rather than being a weakness, such malleability is also thus a survival tactic.

Understanding beyond Models

Dropping allegiance to our commonsense forms of imagined self-exploration means radically changing our perspectives. This means becoming more discerning about our own reasoning processes and more wary of unquestioningly trusting them. This also involves holding every model—every concept—lightly, knowing that it can never accurately reflect or represent reality.

In psychology, for example, this might mean discarding, or changing, treatment interventions that are based on mistaken commonsense beliefs about ourselves.

We must also always remember that all models are mentally constructed. They reflect the way we examine the world, as much as reality itself. These models—my own included—fall short. At best they can assist us; at worst, they can badly mislead us. We should never mistake them for reality—or prefer them over it.

Breaking Free from Mental Constraints

Already, many philosophers encourage people to not cherish any of their constructed models. They believe that nothing less than this is demanded of us if our goal is to know the world as it is. Such flexibility is critical in a changing world, rather than comforting attachments to old ways that blind us to unique and unforeseen events.[4] And therefore, unquestioning faith in anything that doesn't change can be destructive and leave us unprepared for an unpredictable future.

As we know, one model that is ripe for change is the idea of a lasting "self" inside the human body. We have seen that it is only a story that we create. By challenging this inclination to portray humans as relatively fixed entities, we see ourselves as interdependent parts of our communities. We might even view other people as literal extensions of ourselves. Widely adopting this perspective could lead to more of a focus on a spirit of collective evolution rather than individual success.

Fluid Identities and Collective Progress

Also, sometimes for practical purposes we tend to segment societies, perhaps by country, race, sexual preference, gender, and more. But this way of thinking can entangle and limit us, often to the point that we can't see beyond our imaginary divided world.

But by stepping back from our ingrained categories, and perhaps even considering "no-model" thinking, we move beyond our previous illusions and conceptions of ourselves. Hopefully this would help diverse groups progress beyond more divisive concepts, and instead focus on unity, collaboration, and mutual respect.

The inclination to consider no-model thinking, *with the flexibility it requires*, can also make our communities more resilient to shifts in such things as technology, politics, and the environment. With this, communities would be better poised to weather such changes as unified fronts, not as assemblages of individuals.

In psychology, such a shift might help us enormously—and free us from a wide range of mental wild-goose chases and snipe hunts. We would no longer strive to explore unseen imagined conflicts and secret needs. We could waste less time guessing about what supposedly goes on inside our own heads.

The Grace of Not Knowing

As we saw with Anna's journey, hypnosis was a relevant metaphor. In part, it showed that sometimes we don't need to learn complex theoretical models of the mind to make progress. In some cases, we only need to let go of stifling narratives that no longer serve us.

Making such shifts in community thought will require humility. We need to realize and accept that our brains' interpretive processes sometimes greatly impede our perspectives. We need to acknowledge—to ourselves and to each other—that reality is infinitely larger and more complex than any of our models.

Thomas Edison—not a man usually known for modesty—embraced some of this humility when he explained, "We don't know a millionth of one percent about anything."[5] He wasn't kidding or exaggerating—and he wasn't wrong about our collective ignorance.

Seeking new knowledge forces us to acknowledge our limited understandings. It also encourages us to expand them. Perhaps this is like sailing on an endless ocean. The more we explore, the more we see how vast it is.

This attitude shows a willingness to question, challenge, investigate, and, if appropriate, get rid of anything we think we know if it doesn't hold up to scrutiny. It also means recognizing how often common sense can lead us astray.

When it comes to the brain, we can also follow the lead of scientists in physics, biology, chemistry, astronomy, psychology, and many other fields who have achieved major breakthroughs by testing, challenging, investigating, and sometimes disproving widely accepted commonsense models. (In psychology, Pavlov, Freud, and many others engaged in this kind of inquiry. Later pioneers in the field demonstrated similar flexibility by questioning their predecessors' work.)

Unfortunately, because our brains evolved with a priority on meeting self-centered challenges, such a radical shift in our perspective can be very difficult for many of us to make. On top of this, most of us tend to seek evidence that supports our own models—and our own self-centered viewpoints—rather than genuinely investigating reality. This further entrenches our existing commonsense thinking—and allows us to remain unaware of our own biases.

Our biology and our cultures have given us spectacles through which we often misperceive the world. But we also have the power, at least to some extent, to remove those spectacles, see clearly, and choose a different direction.

Any new direction we choose won't be intuitively obvious. It might be like the change physicists faced when they transitioned from somewhat more of a now-commonsense Newtonian physics to relativity and quantum mechanics. It could be as radically different from our habitual way of thinking as coming to understand that colors are partly behavior—or recognizing that there might be no such thing as an unconscious thought.*

Perhaps we will end up shocking ourselves—or shocking ourselves awake. But through this process, those of us who can change our perspectives might discover that we humans are very different creatures than we have long imagined ourselves to be.

---

* Indeed, an "unconscious thought" could be both a fabrication and a contradiction in terms, like a silent B-flat or an invisible glare.

NOTES

Chapter One: Anna and Her Therapist

1. Shaw, S. (2006). *The Jatakas: Birth stories of Bodhisatta*. Penguin.

Chapter Two: Reflective Practitioners

1. Luke 18:1–18.
2. Cooper, A. (2021). *Family-centered support through storytelling: A children's book for hypoplastic left heart syndrome* [Doctoral capstone project portfolio, OCTH 8250]. St. Catherine University. https://sophia.stkate.edu/cgi/viewcontent.cgi?article=1033&context=otd_projects
3. American Psychological Association. (2021). *Guidelines for psychological practice with sexual minority persons*.

Chapter Three: Models Guide and Mislead

1. Dawkins, R. (2006). *The selfish gene*. Oxford University Press.
2. Spanos, N. P. (1996). *Multiple identities and false memories: A sociocognitive perspective* (Ch. 14). American Psychological Association.
3. Hornsey, M. J., Harris, E. A., Bain, P. G., & Fielding, K. S. (2016). Meta-analyses of the determinants and outcomes of belief in climate change. *Nature Climate Change, 6,* 622–626.
4. Skinner, B. F. (1953). *Science and human behavior*. Free Press.
5. Grothaus, M. (2019, July 8). *Bill Gates thinks Steve Jobs was a wizard*. Fast Company. https://www.fastcompany.com/90373250/bill-gates-thinks-steve-jobs-was-a-wizard
6. Aggarwal, A. (2010). The evolving relationship between surgery and medicine. *American Medical Association Journal of Ethics, 12*(2), 119–123; Nolen-Hoeksema, S. (2014). *Abnormal Psychology* (6th ed.). McGraw-Hill Education.
7. Van Swieten, B. (1776). *Commentaries upon Boerhaave's aphorisms concerning the knowledge and cure of disease* (Vol. 11). John Murray.

8. Pradelle, A., Mainbourg, S., Provencher, S., Massy, E., Grenet, G., & Lega, J. (2024). Deaths induced by compassionate use of hydroxychloroquine during the first COVID-19 wave: An estimate. *Biomedicine & Pharmacotherapy, 171*, 116055.

9. Conrad, S. (2012, October 1). Enlightenment in global history: A historiographical critique. *American Historical Review, 117*(4), 999–1027. https://doi.org/10.1093/ahr/117.4.999

10. Botvinick, M., & Cohen, J. (1998). Rubber hand "feels" what eyes see. *Nature, 391*, 756.

11. Niebauer, C. (2019). *No self, no problem: How neuropsychology is catching up to Buddhism.* Hierophant Publishing.

12. Freud, S. (1923). *The ego and the id* (J. Riviere, Trans.). Hogarth Press and Institute of Psychoanalysis.

13. Wihbey, J., Joseph, K., & Lazer, D. (2018). The social silos of journalism? Twitter, news media and partisan segregation. *News Media and Society, 21*(4). https://doi .org/10.1177/1461444818807133

14. Janet, P. (1973). *L'automatisme psychologique: Essai de psychologie experimentale sur les formes inférieures de l'activité humaine* [Psychological automatism: An essay of experimental psychology on the inferior types of human activity]. Société Pierre Janet. (Original work published 1889.)

15. Thigpen, C. H., & Cleckley, H. M. (1957). *The three faces of eve.* Fawcett.

16. Schreiber, F. (1973). *Sybil.* Warner.

17. Spanos, N. P. (1996). *Multiple identities and false memories: A sociocognitive perspective.* American Psychological Association.

18. Smith-Rosenberg, C. (1972). The hysterical woman: Sex roles and role conflict in 19th century America. *Social Research: An International Quarterly, 39*(4), 652–678.

19. Spanos, N. P. (1996). *Multiple identities and false memories: A sociocognitive perspective* (Ch. 15). American Psychological Association.

20. Pigliucci, M. (2009, June 29). In Begley, Sharon. Why do we rape, kill and sleep around? *Newsweek.*

21. American Psychological Association. (2015). Guidelines for psychological practice with transgender and gender nonconforming people. *American Psychologist, 70*(9), 832–864. https://doi.org/10.1037/a0039906

22. Lee, H. (2010). *To kill a mockingbird.* Arrow Books.

Chapter Four: Contemporary Commonsense Models of How the Mind Works

1. Ermer, E., Guerin, S. A., Cosmides, L., Tooby, J., & Miller, M. B. (2006). Theory of mind broad and narrow: Reasoning about social exchange engages ToM areas, precautionary reasoning does not. *Social Neuroscience, 1*(3–4), 196–219.

2. Leslie, A. M. (1987). Pretense and representation: The origins of "theory of mind." *Psychological Review, 94*(4), 412–426.

3. Premack, D., and Woodruff, G. (1978). Does the chimpanzee have a theory of mind? *Behavioral and Brain Sciences, 4*, 515–526.

4. Shapiro, L., & Spaulding, S. Embodied cognition. In Edward N. Zalta (Ed.), *The Stanford Encyclopedia of Philosophy* (Winter 2021 ed.). https://plato.stanford.edu/archives/win2021/entries/embodied-cognition

5. Churchland, P. M. (1981). Eliminative materialism and the propositional attitudes. *Journal of Philosophy, 78*(2), 67–90.

6. Dyer, A. G., Neumeyer, C., & Chittka, L. (2005). Honeybee (*Apis mellifera*) vision can discriminate between and recognise images of human faces. *Journal of Experimental Biology, 208*(24), 4709–4714. https://doi.org/10.1242/jeb.01929

7. Trivers, R. (2000). The elements of a scientific theory of self-deception. *Evolutionary Perspectives on Human Reproductive Behavior, 907*, 114–131.

8. Mele, A. R. (1997). Real self-deception. *Behavioral and Brain Sciences, 20*(1), 91–136; Mele, A. R. (1999). Twisted self-deception. *Philosophical Psychology, 12*(2), 117–137.

9. Wegner, D. M. (2002). *The illusion of conscious will*. MIT Press.

10. Davies, J. M., & Frawley, M. G. (1994). *Treating the adult survivor of childhood sexual abuse*. Basic Books.

Chapter Five: Science and Models of the Mind

1. Bacon, F. (1620). *Novum organum* (Part 2 of *The great instauration*).

2. Watson, J. B. (1919). *Psychology from the standpoint of a behaviorist*. J. B. Lippincott Company.

3. Michelson, A. A., & Morley, E. W. (1887). On the relative motion of the earth and the luminiferous ether. *American Journal of Science, 34*(203), 333–345.

4. Einstein, A., & Infeld, L. (1966). *The evolution of physics: From early concepts to relativity and quanta* (p. 31). Simon and Schuster. (Original work published 1938.)

Chapter Six: The Myth That Our Minds Rarely Make Things Up

1. Gazzaniga, M. S. (2000). Cerebral specialization and interhemispheric communication: Does the corpus callosum enable the human condition? *Brain, 123*, 1293–1326.

2. Gazzaniga, M. S. (2000). Cerebral specialization and interhemispheric communication: Does the corpus callosum enable the human condition? *Brain, 123*, 1293–1326.

3. Gazzaniga, M. S. (2000). Cerebral specialization and interhemispheric communication: Does the corpus callosum enable the human condition? *Brain, 123*, 1293–1326.

4. Gazzaniga, M. S. (2000). Cerebral specialization and interhemispheric communication: Does the corpus callosum enable the human condition? *Brain, 123*, 1293–1326.

5. Gazzaniga, M. S., LeDoux, J. E., & Wilson, D. H. (1977). Language, praxis, and the right hemisphere: Clues to some mechanisms of consciousness. *Neurology, 27*, 1144–1147.

6. Gazzaniga, M. S. (1995). Consciousness and the cerebral hemispheres. In Michael S. Gazzaniga (Ed.), *The cognitive neurosciences*. MIT Press.

7. Gerstmann, J. (1942). Problem of imperception of disease and of impaired body territories with organic lesions: Relation to body scheme and its disorders. *Archive of Neurology and Psychiatry, 48*, 890–913.

8. Gerstmann, J. (1942). Problem of imperception of disease and of impaired body territories with organic lesions: Relation to body scheme and its disorders. *Archive of Neurology and Psychiatry, 48*, 890–913.

9. Conchiglia, G., Della Rocca, G., & Grossi, D. (2007). On a peculiar environmental tendency syndrome in a case with frontal-temporal damage: Zelig-like syndrome. *Neurocase, 13*(1), 1–5.

10. Loftus, E. (2003). Make-believe memories. *American Psychologist, 58*(11), 867–873.

11. Ceci, S. J., Huffman, M. L. C., Smith, E., & Loftus, E. F. (1994). Repeatedly thinking about a non-event: Source misattributions among preschoolers. *Consciousness and Cognition, 3*, 388–407.

12. Nisbett, R. E., and Wilson, T. D. (1977). Telling more than we can know: Verbal reports on mental processes. *Psychological Review, 84*(3), 231–259.

13. Meehl, P. E. (1954). *Clinical vs. statistical prediction: A theoretical analysis and a review of the evidence*. University of Minnesota Press.

14. Gould, S. J. (1994). In the mind of the beholder. *Natural History, 103*(2), 14–23.

Chapter Seven: The Myths and Making of the Self

1. James, W. (1890). *The Principles of psychology*. Henry Holt and Company.

2. Goldman, B. (2023, June 22). Sense of self: The brain structure that holds key to "I." Stanford University.

3. Herwig, U. (2010). Me, myself and I. *Scientific American Mind, 21*(3), 58–63.

4. Chiao, J. Y., Harada, T., Komeda, H., Li, Z., Mano, Y., Saito, D., Parrish, T. B., Sadato, N., & Iidaka, T. (2009). Neural basis of individualistic and collectivistic views of self. *Human Brain Mapping, 30*(9), 2813–2820.

5. Yong, J. C., Li, N. P., & Kanazawa, S. (2021). Not so much rational but rationalizing: Humans evolved as coherence-seeking, fiction-making animals. *American Psychologist, 76*(5), 781–793. https://doi.org/10.1037/amp0000674

Chapter Eight: The Limits of a Relatively Fixed Self in a Dynamic World

1. Fast, I. (1998). *Selving: A relational theory of self-organization* (p. 6). Analytic Press.

2. Hoijer, H. (Ed.). (1954). *Language in culture: Conference on the interrelations of language and other aspects of culture*. University of Chicago Press.

3. Borges, J. L. (1983). "Tlön, Uqbar, Orbis Tertius." Porcupine's Quill.

Chapter Nine: The Promises of a Fluid Self

1. Tedeschi, R. G., & Calhoun, L. G. (2004). Posttraumatic growth: Conceptual foundations and empirical evidence. *Psychological Inquiry, 15*(1), 1–18.

2. Frankl, V. E. (1959). *Man's search for meaning: An introduction to logotherapy.* Beacon Press.

Chapter Ten: The Myth of an Unconstructed Reality

1. Kohler, I. (1961). Experiments with goggles. *Scientific American, 206,* 62–72.

2. Kanazawa, S. (2010). Evolutionary psychology and intelligence research. *American Psychologist, 65*(4), 279–289.

3. Lanza, R. L., Pavšič, M., & Berman, B. (2020). *The grand biocentric design: How life creates reality.* BenBella Books.

4. Sheng-yen. (1990). In The twelve entries [Lecture given on the Śūraṅgama Sūtra]. *Ch'an Newsletter, 81.*

5. Hume, D. (1739). *A treatise of human nature: An attempt to introduce the experimental method of reasoning into moral subjects.* John Noon.

6. Cordova, V. F. (2007). *How it is: The Native American philosophy of V. F. Cordova* (K. D. Moore, K. Peters, T. Jojola, & A. Lacy, Eds.). University of Arizona Press.

Chapter Eleven: The Myth That We Can Peer Inside Our Own Heads

1. James, W. (1890). *Principles of psychology.* Henry Holt and Company.

2. Sellars, W. (1956). Empiricism and the philosophy of mind. In H. Feigl and M. Scriven (Eds.), *The foundations of science and the concepts of psychology and psychoanalysis: Minnesota studies in the philosophy of science* (vol. 1). University of Minnesota Press.

Chapter Twelve: The Myth That Hypnosis Is a Unique State of Mind

1. Kosslyn, S. M., Thompson, W. L., Constantini-Ferrando, M. F., Alpert, N. M., & Spiegel, D. (2000). Hypnotic visual illusion alters color processing in the brain. *American Journal of Psychiatry, 157,* 1279–1284.

2. Scheibe, K. E., Gray, A. L., & Keim, C. S. (1968). Hypnotically induced deafness and delayed auditory feedback: A comparison of real and simulating subjects. *International Journal of Clinical and Experimental Hypnosis, 16,* 158–164.

3. Darwin, C. (1859). *On the origin of species by means of natural selection.* John Murray.

4. Richerson, P. J., & Boyd, R. (2005). *Not by genes alone: How culture transforms human evolution.* University of Chicago Press.

5. Krack, R. (2006). *Thaipusam: On needles and pins in Singapore.* Rainer Krack/CPAmedia.

6. Pinker, S. (1997). *How the mind works*. Penguin.

7. Kirsch, I., & Lynn, S. J. (1997). Hypnotic involuntariness and the automaticity of everyday life. *American Journal of Clinical Hypnosis, 40*(1), 329–348.

8. Flemons, D. (2020). Toward a relational theory of hypnosis. *American Journal of Clinical Hypnosis, 62*(4), 344–363. https://doi.org/10.1080/00029157.2019.1666700

Chapter Thirteen: Hypnosis and Our Interconnected Reality

1. Ireland Weir, K. (2024, April/May). Uncovering the new science of clinical hypnosis. *Monitor on Psychology, 55*(3), 47.

2. Whitman, W. (1897/1914). *Leaves of grass.* New York: Mitchell Kennerley.

3. Hsüan Hua and Buddhist Text Translation Society (Eds.). (2009). *The Śūraṅgama Sūtra: A new translation.* Buddhist Text Translation Society.

4. Allen, S. (2018, September). The science of awe [White paper prepared for the John Templeton Foundation]. Greater Good Science Center at UC Berkeley.

5. Allen, S. (2018, September). The science of awe [White paper prepared for the John Templeton Foundation]. Greater Good Science Center at UC Berkeley.

Chapter Fourteen: The Myth of a Healthy and Single "I"

1. American Psychiatric Association. (2013). *Diagnostic and statistical manual of mental disorders* (5th ed.).

2. Flemons, D. (2020). Toward a relational theory of hypnosis. *American Journal of Clinical Hypnosis, 62*(4), 344–363. https://doi.org/10.1080/00029157.2019.1666700

3. Watkins, H. H. (1993). Ego-state therapy: An overview. *American Journal of Clinical Hypnosis, 35*(4), 232–240. https://doi.org/10.1080/00029157.1993.10403014. See also Watkins, J. G., & Watkins, H. H. (1988). The management of malevolent ego states in multiple personality disorder. *Dissociation, 1,* 1.

Chapter Fifteen: The Myths of Conscious Control and Self-Understanding

1. Melnick, M. D., Tadin, D., & Huxlin, K. R. (2016). Re-learning to see in cortical blindness. *Neuroscientist, 22*(2), 199–212. https://doi.org/10.1177/1073858415621035

2. James, W. (1890). *The principles of psychology.* Henry Holt and Company.

3. Skinner, B. F. (1953). *Science and human behavior* (p. 279). Free Press.

4. Pinker, S. (1977). *How the mind works* (Preface). Penguin.

5. Pinker, S. (1997). *How the mind works*. Penguin.

6. Pinker, S. (1997). *How the mind works*. Penguin.

7. Libet, B. (1985). Unconscious cerebral initiative and the role of conscious will in voluntary action. *Behavioral and Brain Sciences, 8*, 529–566.

8. Soon, C. S., Brass, M., Heinze, H.-J., & Haynes, J.-D. (2008, April 13). Unconscious determinants of free decisions in the human brain (Abstract). *Nature Neuroscience, 11*(5), 543–545.

9. Freeman, W. J. (2001, May 31–June 2). Bridging the gaps between neuron, brain and behavior with neurodynamics [Jean Piaget Society symposium]. Berkeley, CA.

Chapter Sixteen: Autonomous Self-Control

1. Wegner, D. M., & Wheatley, T. (1999). Apparent mental causation: Sources of the experience of will. *American Psychologist, 54*(7), 480–492.

2. Harari, Y. N. (2017). *Homo deus: A brief history of tomorrow*. Harper Perennial.

3. Smullyan, R. M. (1977). Is God a Taoist? In D. R. Hofstadter and D. C. Dennett (Eds.), *The Mind's I* (1981). Basic Books.

4. Metzinger, T. (2009). *The ego tunnel*. Basic Books.

Chapter Seventeen: Human Consciousness and Inner Peace

1. Sutherland, N. S. (Ed.). (1989). *The international dictionary of psychology*. Continuum.

2. Chalmers, D. J. (1996). *The conscious mind: In search of a fundamental theory* (p. 6). Oxford University Press.

3. Hagen, S. (2020). *The grand delusion*. Wisdom Publications.

4. Leiber, J. (1996). Helen Keller as cognitive scientist. *Philosophical Psychology, 9*(4), 419–440.

5. Keller, H. (1904/1908). *The world I live in*. Century.

6. Yu Family School. (1127–1279). *Dao de jing*. Jian'an, Fujian.

7. Burns, T. R., & Engdahl, E. (1998). The social construction of consciousness: Individual selves, self-awareness, and reflectivity. *Journal of Consciousness Studies, 5*(2), 166–184.

Chapter Eighteen: Generosity beyond Genetics

1. Klimecki, O. M., Leiberg, S., Lamm, C., & Singer, T. (2013). Functional neural plasticity and associated changes in positive affect after compassion training. *Cerebral Cortex, 23*(7), 1552–1561.

2. Rizzolatti, G., & Craighero, L. (2004). The mirror-neuron system. *Annual Review of Neuroscience, 27*, 169–192.

3. Dawkins, R. (2006). *The selfish gene*. Oxford University Press.

4. Abramson, A. (2021, November). Cultivating empathy. *Monitor on Psychology, 52*(8).

5. Blanken, I., van de Ven, N., & Zeelenberg, M. (2015). A meta-analytic review of moral licensing. *Personality and Social Psychology Bulletin, 41*(4). https://doi.org/10.1177/0146167215572134

6. Weng, H. Y., Fox, A. S., Shackman, A. J., Stodola, D. E., Caldwell, J. Z. K., Olson, M. C., Rogers, G. M., & Davidson, R. J. (2013). Compassion training alters altruism and neural responses to suffering. *Psychological Science, 24*(7), 1171–1180. https://doi.org/10.1177/0956797612469537

Chapter Nineteen: Self, Culture, Artificial Intelligence, and Philosophy

1. Wu, J., & Rao, R. P. N. (2017, April 9). *Melding mind and machine: How close are we?* The Conversation. https://theconversation.com/melding-mind-and-machine-how-close-are-we-75589?xid=PS_smithsonian

2. Wu wei. (1994). *I Ching wisdom: Guidance from the Book of Changes.* Power Press.

3. Song, B. (2020, August 20). Applying ancient Chinese philosophy to artificial intelligence. *Noema Magazine.* https://www.noemamag.com/applying-ancient-chinese-philosophy-to-artificial-intelligence

4. Song, B. (2023). How Chinese philosophy impacts AI narratives and imagined AI futures. In S. Cave & K. Dihal (Eds.), *Imagining AI: How the World Sees Intelligent Machines* (pp. 338–352). Oxford University Press.

5. Song, B. (2023). How Chinese philosophy impacts AI narratives and imagined AI futures. In S. Cave & K. Dihal (Eds.), *Imagining AI: How the World Sees Intelligent Machines* (pp. 338–352). Oxford University Press.

6. Johnson, C., & Tyson, A. (2020, December 15). *People globally offer mixed views of the impact of artificial intelligence, job automation on society.* Pew Research Center.

7. Gould, H., & Walters, H. (2020). Bad Buddhists, good robots: Techno-salvationist designs for Nirvana. *Journal of Global Buddhism, 21*, 277–294.

8. Gould, H., & Walters, H. (2020). Bad Buddhists, good robots: Techno-salvationist designs for Nirvana. *Journal of Global Buddhism, 21*, 277–294.

9. Gould, H., & Walters, H. (2020). Bad Buddhists, good robots: Techno-salvationist designs for Nirvana. *Journal of Global Buddhism, 21*, 277–294.

10. Gould, H., & Walters, H. (2020). Bad Buddhists, good robots: Techno-salvationist designs for Nirvana. *Journal of Global Buddhism, 21*, 277–294.

Chapter Twenty: Embracing a Unified Reality

1. Jones, R. S. (1992). *Physics for the rest of us: Ten basic ideas of twentieth-century physics that everyone should know . . . and how they have shaped our culture and consciousness* (p. 221). Contemporary Books.

2. Siegel, D. J. (2017). *Mind: A journey to the heart of being human.* Norton.

3. Mulligan, R. (Director). (1962). *To Kill a Mockingbird* [Film]. Universal Pictures.

Chapter Twenty-One: Truth, Humility, and a World beyond Models

1. Batchelor, S. (1997). *Buddhism without beliefs*. Penguin.

2. Shiah, Y.-J. (2016). From self to nonself: The nonself theory. *Frontiers in Psychology, 7*. https://doi.org/10.3389/fpsyg.2016.00124

3. Shiah, Y.-J. (2016). Theoretical and philosophical psychology. *Frontiers in Psychology, 7*. https://doi.org/10.3389/fpsyg.2016.00124

4. Batchelor, S. (1997). *Buddhism without beliefs*. Penguin.

5. Edison, T. (1931). As reported in B. E. Stevenson, *Stevenson's Book of Quotations* (3rd ed.). Cassell.

ACKNOWLEDGMENTS

An exceptional group of individuals has given me the insights necessary to complete this text. Scott Edelstein's professionalism and remarkable ability to weave together diverse concepts have been invaluable. Dr. Michael Schulein's candor and scholarship have, for many years, challenged my thinking. Drs. Michael Nash, Peter Zelles, Roger Jones, Joy Laine, Douglas Flemons, Susan McPherson, and Jonathan Stoltz offered learned appraisals of the manuscript. Steve Hagen, Sean Murphy, and Ben Connelly, each one an author and teacher of science, philosophy, and Eastern thought, provided valuable insights. Senior pastor Jeff Lindsay and Dr. John Froula, associate professor of dogmatic theology, offered helpful comments from perspectives different from my own. Anita Fisher's insightful reviews, and decades-long endurance of my many quirks (including my years of sudden disappearances to record manuscript ideas), reveal her unwavering support for me and for this project. I am grateful to my parents, Charles and LaVon, who, during my childhood, encouraged me to believe that nothing was beyond the scope of questioning. Ralph Lindell's global outlook has broadened my views, and Daniel Pichelman's technical thinking has helped and challenged my own thoughts.

I hope that the mutual journey I have taken with each of these people—and others—will inspire curiosity in many more.

ABOUT THE AUTHOR

David C. Fisher, Ph.D., is a psychologist emeritus, a diplomate in clinical psychology, a fellow of the American Psychological Association (APA), and a recipient of APA's Innovative Practice Award. Dr. Fisher's exploration of Eastern thinking began in the 1970s, a pursuit that greatly enhanced his knowledge of human thought. In his spare time, you might find him painting portraits or enjoying the serenity of a tranquil lakeside.